INTRODUCTION TO WLLs

IEEE Press Series on Digital & Mobile Communication

The IEEE Press Digital and Mobile Communication Series is written for research and development engineers and graduate students in communication engineering. The burgeoning wireless and personal communication fields receive special emphasis. Books are of two types, graduate texts and the latest monographs about theory and practice.

Books in the IEEE Press Series on Digital & Mobile Communication

John B. Anderson, *Digital Transmission Engineering*
Rolf Johannesson and Kamil Sh. Zigangirov, *Fundamentals of Convolutional Coding*
Raj Pandya, *Mobile and Personal Communication Systems and Services*
Lajos Hanzo, P. J. Cherriman, and J. Streit, *Video Compression & Communications over Wireless Channels: Second to Third Generation Systems and Beyond*
Lajos Hanzo, F. Clare, A. Somerville and Jason P. Woodard, *Voice Compression and Communications: Principles and Applications for Fixed and Wireless Channels*
Mansoor Shafi, Shigeaki Ogose and Takeshi Hattori (Editors), *Wireless Communications in the 21st Century*
Raj Pandya, *Introduction to WLLs: Application and Deployment for Fixed and Broadband Services*

INTRODUCTION TO WLLs

APPLICATION AND DEPLOYMENT FOR FIXED AND BROADBAND SERVICES

Raj Pandya

John B. Anderson, *Series Editor*

IEEE PRESS

A John Wiley & Sons, Inc., Publication

IEEE Press
445 Hoes Lane,
Piscataway, NJ 08855

Published simultaneously in Canada.

Library of Congress Cataloging-in-Publication Data:

Pandya, Raj, 1932–
Introduction to WLLS : application and deployment for fixed and broadband services/
Raj Pandya.
p. cm.
Includes bibliographical references and index.
ISBN 0-471-45132-0 (cloth)

1. Wireless communication systems. 2. Local loop (Telephony) I. Title.

TK5103.2.P35 2003
621.382–dc22 2003057639

Printed in the United States of America

10 9 8 7 6 5 4 3 2 1

To the International Telecommunications Union for Its Leadership in Promoting Telecommunication Services Around the World

CONTENTS

PREFACE

The motivation for this book originated from my recent ITU/UNDP assignment in India as a Senior Expert on wireless local loop (WLL) technology. The primary purpose of the assignment was to develop a short course on the fundamentals of WLL systems at the Telecommunications Training Center of the Indian Telecommunications Corporation (Bharat Sanchar Nigam Ltd.). The course was intended for the training of corporate engineers and technical managers who would be involved in the planning, implementation, and operation of WLL systems that are being planned and deployed rapidly throughout India. It was disappointing that no suitable reference text on WLL systems was available that I could recommend as a supplement and follow-up reading for the short course. The scope and style of this book has been tailored to fill this gap and to provide a technical reference on WLL systems for telecommunication engineers and managers engaged in the planning, design, and operation of wireless access networks around the world.

Wireless local loop (WLL) is now widely recognized as an economically viable technology for provision of telecommunication services to subscribers in sparsely populated as well as highly congested areas because of its many advantages over its wired counterpart. However, the preparation of the business case, choice of a suitable technology, deployment planning, and radio and network system design for a WLL system depend on a range of technical and strategic planning variables. In order to successfully manage a WLL system from initial concept to final implementation and operation, the engineers and technical managers need to have an appreciation of the range of technical and planning issues associated with WLL systems.

The primary objective of this book is to provide a systems level view of the technical, planning, and deployment aspects of WLL systems, as an alternative technology to a copper-based local loop plant for the provision of basic (narrowband) telecommunication services such as voice and voice-band data to end users. However, this book also includes an overview of broadband wireless access (BWA) systems that are emerging as an important alternative to fiber, cable modem, and copper (digital subscriber loop or DSL) systems to meet the increasing demand for such services as video on demand, Internet access, and high-speed data services.

A major part of the book is devoted to the background for and discussion of WLL systems, primarily because of the large worldwide market for WLL systems and the resulting need for suitable reference text material on the topic. However, since a significant market is emerging for BWA systems, this book includes a chapter that addresses multipoint distribution, wireless LAN, and satellite-based technologies and systems for providing broadband services. For pedagogical reasons, an effort has been made to keep the discussion of wireless access for narrowband services (WLL) and broadband services (BWA) separate and distinct, though some of the basic principles of radio systems equally apply to both aspects of wireless access.

The preparation of the business case, choice of a suitable technology, deployment planning, and radio and network system design for a WLL system depend on a range of technical and strategic planning variables. In order to successfully manage a WLL system from initial concept to final implementation and operation, the engineers and technical managers need to have an appreciation of the range of technical and planning issues associated with WLL systems. This book addresses the basic principles, technologies, and planning, and deployment aspects of WLL systems, and it provides some examples of commercially available WLL systems and the technology choices they offer. It is hoped that the book will be a useful guide to engineers and managers who may be involved in the planning, design, and operation of access networks.

In terms of the depth and breadth of coverage, this book is targeted to serve the following audiences and their needs:

- As a technical training text and guide for telecommunication engineers and scientists who expect to engage in planning and procurement of WLL and BWA systems for network operators and telecommunications service providers.
- As a reference text for technical managers who are engineering and marketing WLL and BWA systems.
- As a reference text for senior telecommunications engineering students at the university and technical college level.

This book consists of eight chapters followed by a Bibliography and a Glossary of Terms.

Chapter 1 is the introductory chapter, which covers the background material including terminology of wireless access systems, history and forecasts for worldwide WLL deployment, advantages and disadvantages of wireless access, comparison between WLL and cellular mobile systems, and progress in international and regional standards development organizations such as ITU and ETSI.

Chapter 2 provides some background on fundamental aspects of radio systems relevant for wireless access systems including such topics as radio spectrum and frequency bands; classification of radio systems; principles of duplexing and multiple access schemes; source coding, channel coding, and interleaving; and basic and high-bit-rate modulation methods.

Chapter 3 provides a systems level description of digital cellular mobile and cordless telecommunication systems that provide technical underpinnings for many commercial WLL systems. The systems described include: GSM (European TDMA system), TIA/EIA IS 136 (North American TDMA system), TIA/EIA IS 95 (North American CDMA system), DECT (European cordless telecommunication system), and PHS (Japanese telecommunication cordless system). This chapter also provides an overview of enhancements to TDMA and CDMA cellular mobile systems to support high-bit-rate data services and their evolution toward third-generation (3G) mobile communication systems.

Chapter 4 covers the system components and interfaces that are generally associated with a WLL system and includes (a) the descriptions of basic system components such as indoor unit and outdoor unit and (b) the range of radio interfaces and network interfaces including the V5.2 interface commonly deployed in WLL systems.

Chapter 5 addresses the radio design aspects of WLL systems which cover the following: radio propagation characteristics; radio path loss and radio link availability; frequency planning and frequency reuse for WLL systems; radio network planning aspects; and comparison of WLL radio design with cellular mobile systems.

Chapter 6 provides a description of the necessary steps in the planning and deployment of a WLL system that need to be followed by a prospective WLL network operator. The procedures described include: assessing the service needs including the effects of supporting limited or full mobility; estimating the traffic load and its distribution; choosing a suitable WLL technology; and final radio engineering and network design.

Chapter 7 is intended to describe a sample of commercially available WLL systems based on existing cellular mobile and cordless telecommunication system standards as well as systems based on proprietary radio technologies.

Chapter 8 provides a comprehensive overview of currently deployed broadband wireless access (BWA) systems such as LMDS (local multipoint distribution system) and MVDS (multipoint video, distribution system) as well as emerging BWA system such as IEEE 802.16, ETSI-BRAN HIPERACCESS, and satellite-based BWA systems.

In a book like this, which addresses a variety of technologies, services, and standards associated with wireless access, frequent use of acronyms and abbreviations is almost inevitable and unavoidable. Many of the readers who have some background in wireless and mobile communication systems are likely to be familiar with many of the acronyms. However, in order to partially alleviate the readers' frustration in dealing with this perennial problem, the author has attempted to ensure that the acronyms are spelled out where they first appear in the text, and an extensive Glossary is provided at the end of the book.

In conclusion, this book represents my effort at developing a reference text on wireless access technologies and systems for basic (narrowband) and broadband telecommunication services which is sufficiently comprehensive in terms of scope, easily comprehensible in terms of style and organization, and useful to a very large audience of engineers and technical managers who are involved in planning and design of telecommunication networks around the world.

Acknowledgments

The support provided by my family, friends, and colleagues toward completion of the book is gratefully acknowledged. Special acknowledgment is also due to the International Telecommunications Union (Development Sector) for the technical assignment in India which provided the initial motivation and material for this book.

RAJ PANDYA

Toronto, Ontario, Canada
September 2003

INTRODUCTION TO WLLs

CHAPTER 1

Introduction to Wireless Access and WLL System Deployment

Wireless technology is now recognized as an important option for delivering mobile, fixed, and broadband services to the end users. Wireless local loop (WLL) utilizes wireless access for last mile/first mile connectivity to the subscriber premises to provide basic telecommunication services. This introductory chapter covers such topics as: related terminology; different access network architectures including wireless access options; advantages and potential limitations of WLL and wireless access; comparison between WLL and cellular mobile systems; and the ongoing standardization activities on fixed and broadband wireless access in international and regional standards development organizations like ITU (International Telecommunications Union), ETSI (European Telecommunications Standards Institute), and IEEE (Institution of Electrical and Electronics Engineering). The intent of this chapter is to provide a background on the applications and economics of wireless access for providing last mile/first mile connectivity to support fixed narrowband and broadband services.

1.1 TERMINOLOGY OF WIRELESS ACCESS SYSTEMS

The term *wireless local loop (WLL)* has been in use for some time and is well-ingrained in the literature. It originated from the initial applications of radio technology in place of a copper local loop plant for providing a last mile/first mile link for delivery of basic telephony services to end users. In recent years, alternate terms have been proposed and are in use to reflect a range of applications for wireless access. The term *radio in the local loop (RLL)* was used by the European Telecommunications Standards Institute (ETSI) in some of their documents describing similar use of radio technology in the access network. The Radio Communication Sector of the International Telecommunications Union (ITU-R), as part of its standardization studies on wireless access, has proposed the classification and terminology for wireless access applications shown in Figure 1.1. Thus, ITU-R

Introduction to WLLs. By Raj Pandya
ISBN 0-471-45132-0

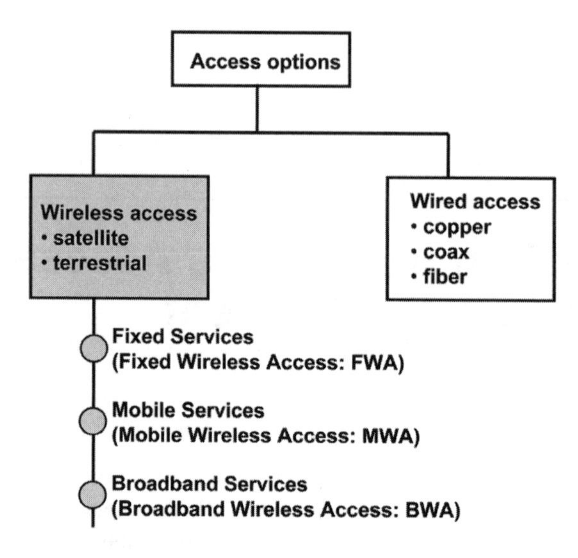

Figure 1.1. ITU-R classification of access types and wireless access services.

has proposed fixed wireless access (FWA) as a somewhat broader term than WLL for terrestrial wireless access systems to provide basic telecommunication services to fixed (as opposed to mobile) terminals/subscribers. However, in order to maintain continuity with the majority of the published literature and minimize reader confusion, the term wireless local loop (WLL) has been used in this book with the term FWA used only on an exception basis. Note that terrestrial as well as satellite technologies are included for provision of wireless access services in the above classification.

1.2 HISTORICAL BACKGROUND AND GROWTH FORECASTS FOR WLL

The use of radio telephony to provide basic communications service to distant, isolated communities in such countries as Canada and Australia has been attempted since the 1950s using single-channel very high frequency (VHF) radios. However, these early attempts were more in the form of a public service (heavily subsidized by the government) rather than a commercial offering on the same level as the wireline telephony service. The user acceptability and large-scale commercial deployment of wireless access systems for provision of telecommunication services in the 1950s and 1960s was hindered by such factors as

- Lack of suitable technologies for sharing frequency spectrum
- High-cost and low-performance of the radio equipment

- Lack of user-friendly operational procedures
- Lack of reliability
- Poor quality of service encountered by the users

The technological advances in wireless access systems to provide mobile services during 1970s and 1980s leading to the analog cellular technology and subsequently (in 1990s) to digital cellular technologies provided the underpinnings for a viable business case for WLL systems. The viability of the business case for WLL systems resulted from the unprecedented demand for cellular mobile and personal communication services which led to large volume production of wireless access equipment and terminal devices with accompanying reduction in their production costs and highly competitive prices.

A key circumstance in boosting the role of WLL for provision of basic telephony service and increasing its profile as a viable alternative to wired access was provided by the aftermath of the unification of East and West Germany. In order to bring the East German telecommunication infrastructure closer to the highly advanced levels in West Germany, in a rapid and cost effective manner, the German government opted for extensive deployment of WLL systems based on the Nordic Mobile Telephone system (NMT) analog cellular technology.

During this period, a major study was undertaken by the European Bank for Reconstruction and Development (EBRD) which looked at the policy options for telecommunications sector in Central and East Europe and the former Soviet Union. This study included extensive comparison of annual life cycle costs of various wireline and wireless access technologies as a function of subscriber densities. The study clearly indicated that, assuming spectrum availability and average residential calling rates, wireless access technologies were economical compared to underground copper for subscriber densities below about 200–400 subscribers/km^2. The actual crossover point will, of course, vary depending on distribution of subscriber densities, traffic demand levels, and the wireless technology deployed.

Since the early 1990s there has been a rapid proliferation of WLL systems based on cellular and cordless radio technologies as well as those based on proprietary radio technologies. These systems are being deployed worldwide. For example, India has started deploying WLL systems based on code division multiple access (CDMA) and digital enhanced cordless telecommunications (DECT) radio standards which not only provide fixed telephony service but also support limited mobility.

Figure 1.2 and Table 1.1 provide worldwide growth estimates for WLL subscribers and for WLL equipment revenues prepared by Frost and Sullivan in 1999 and available at the CDMA Development Group (CDG) website (www.cdg.org). Note that in the light of generally depressed telecommunications markets in the last few years, these estimates may now be some what optimistic in some regions.

As illustrated by Figure 1.2, the major growth in WLL subscribers is expected to take place in Asia and Latin America, where major efforts are underway by many developing countries to improve telecommunications services to their populations in a rapid and cost-effective manner. Table 1.1 shows the revenue forecasts in terms of

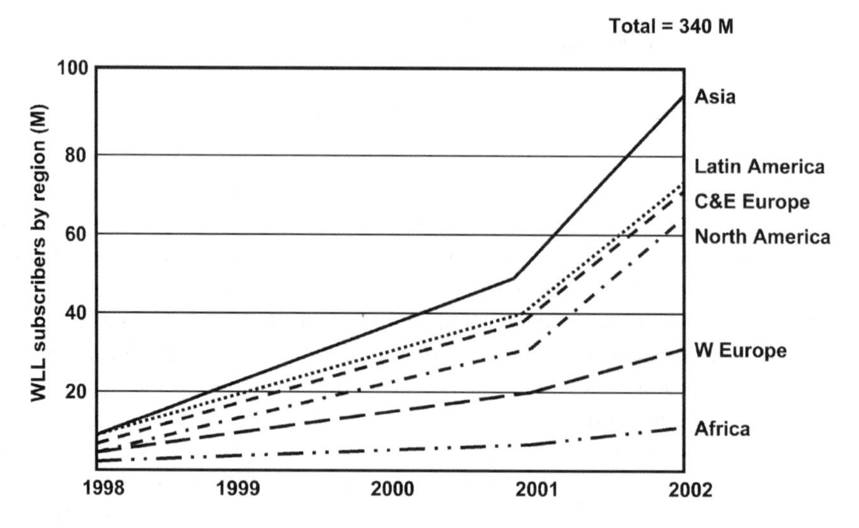

Figure 1.2. Estimates for worldwide WLL subscribers by region. *Source*: CDMA Development Group (www.cdg.org).

WLL terminals, base stations, and backhaul equipment (equipment required to connect WLL base stations to public switched network).

1.3 ACCESS NETWORKS AND WIRELESS ACCESS OPTIONS

Figure 1.3 illustrates the range of access arrangements that are possible for connecting the end-user equipment to the local switching system. This represents the

TABLE 1.1. Revenue Forecasts in $M (1998–2005) for WLL Equipment

Year	Total	CPE	Base Stations	Backhaul Equipment
1998	1,603.0	806.3	649.2	147.5
1999	3,107.3	1,727.7	1,131.1	248.6
2000	5,480.5	3,480.1	1,688.0	312.4
2001	8,510.0	5,863.4	2,306.0	2340.4
2002	12,375.0	9,108.0	2,871.0	396.0
2003	16,211.3	12,158.4	3,566.5	486.3
2004	19,777.7	15,624.4	3,757.8	395.6
2005	23,337.7	18,903.5	3,967.4	466.8
CAGR	39.8%	48.3%	25.4%	15.5%

Note: All figures are rounded, base year is 1999.
Source: CDMA Development Group (www.cdg.org).

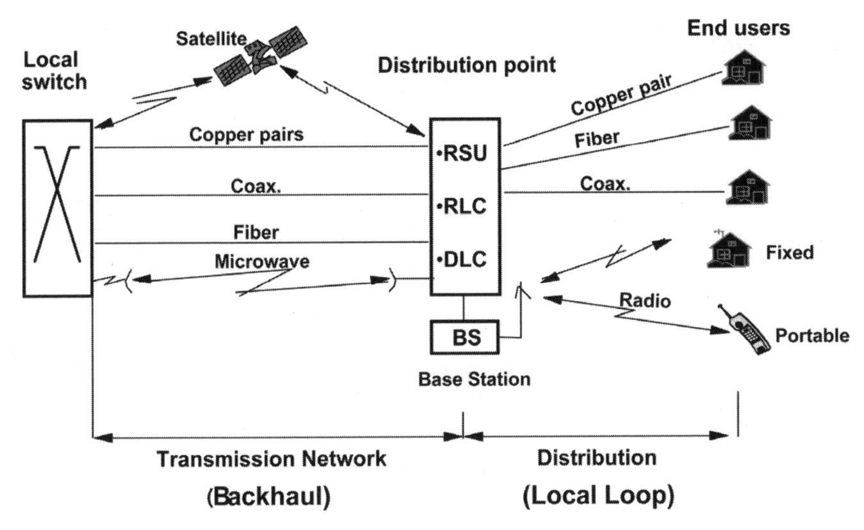

Figure 1.3. Access networks and range of available access technologies.

so-called *external plant* required to deliver switched services to the end users. Generally, the access network has two basic components: (a) the access part, which in the classical terminology constitutes the *local loop*, and (b) the *backhaul* part, which connects the distribution point to the local switch. Depending on the specific application and the choice of access technology, the distribution point may take the form of a remote switching unit (RSU), a remote line concentrator (RLC), a digital line carrier (DLC), or a (radio) base station (BS).

As evident from Figure 1.3, a range of technologies, including wireless, can be deployed in the local loop as well as in the backhaul portions of the access network.

Figure 1.4 further elaborates on the leading access technologies that are being deployed in current networks in order to support various application environments in an economical and efficient manner. Whereas technologies that utilize digital line carriers, remote switching, and remote concentrators represent nonwireless access technologies, there are various options for utilizing wireless for local access. These range from cellular systems, WLL systems, and satellites. Obviously, the choice between different wireless access technologies will depend on the applications/ services that need to be supported and the associated economics. Satellites are deployed for delivery of basic telecommunication services only in very specific cases for providing communication links to distant, isolated locations (e.g., remote oil fields).

Figure 1.5 illustrates the basic architecture for a WLL system. In a WLL system the subscriber access is provided by a multiple access radio system that may operate

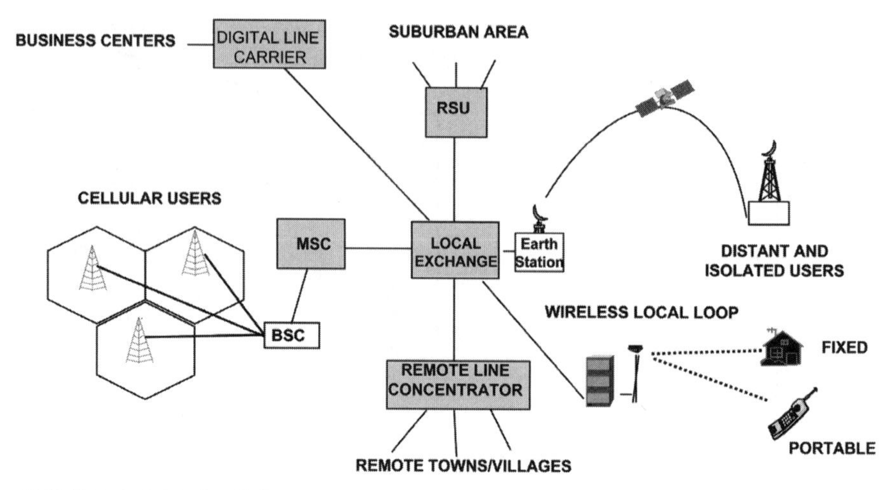

Figure 1.4. Different access types and their application environments.

in a radio-frequency band in the range of 400 MHz to 40 GHz. These multiple access radio systems are designed to maximize the utilization of the available frequency spectrum, which is generally a scarce and highly regulated resource. Unlike the case of a wireline access technology using copper pairs, where each subscriber's premises are provided with an individual connection (always available), the multiple access radio system for WLL requires a sharing of the available radio channels among a large number of subscribers on a demand basis. In other words, the radio access system is designed to provide a specified grade of service (e.g., 1% blocking probability) and the individual subscribers have to compete for the available radio channels. Besides the fact that a voice/data channel is not always guaranteed in the radio system, the channel itself may suffer degraded quality because of the interference and signal loss associated with radio channels.

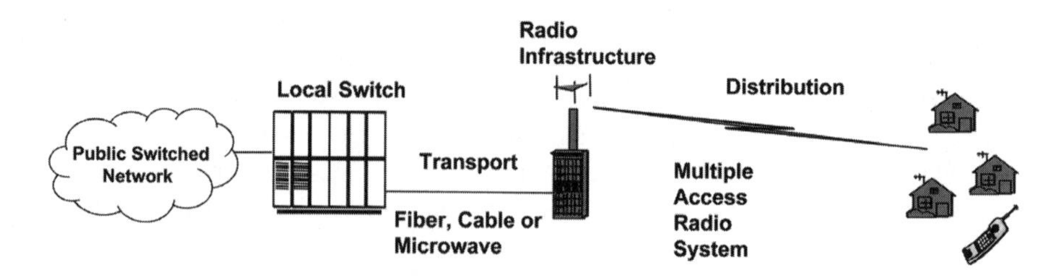

Figure 1.5. Basic architecture typically deployed for WLL systems.

The three multiple access methods commonly used for creating multiple channels from the available frequency spectrum for increased efficiency include frequency division multiple access (FDMA), time division multiple access (TDMA), and code division multiple access (CDMA) methods.

1.4 ADVANTAGES AND ECONOMICS OF WLL SYSTEMS

Inspite of some risk associated with radio channel availability and radio channel quality in WLL systems, there are a number of market forces that are driving the deployment of wireless access (as opposed to wireline access) in various application environments. Some of these forces include the following:

- The tremendous demand for new business and residential telephone services in many developing countries can be met quickly and economically by deploying WLL systems.
- Existing wireline operators can extend their networks with WLL to serve rural, suburban, and urban subscribers.
- Cellular operators can leverage their current mobile networks to deliver local residential services using WLL.
- With the emerging competitive environment for local telephone services in developed and developing countries, new service providers can quickly deploy suitable WLL solutions to provide telecommunication services in a community.

Besides the emerging market forces mentioned above, there are also operational and cost efficiencies that make a wireless access solution more attractive. These are addressed in the next section.

Following are some of the advantages of deploying wireless local loop systems compared to wireline access systems:

- High speed of deployment of the network infrastructure and customer units
- Low initial capital investment in infrastructure
- Installation of network and customer infrastructure (e.g., base stations) only when needed
- Ability to selectively provide services only in areas where and when required
- Potential for reduced operation and maintenance costs
- Less sensitivity to forecast errors
- Allows the provision of limited mobility in addition to fixed services

In terms of cost comparisons between wireless and wireline access systems one needs to consider such cost components as capital costs, operational costs, and opportunity costs associated with the planning, design, deployment, operation, and maintenance of these systems. Figure 1.6 provides a comparison of wireline and

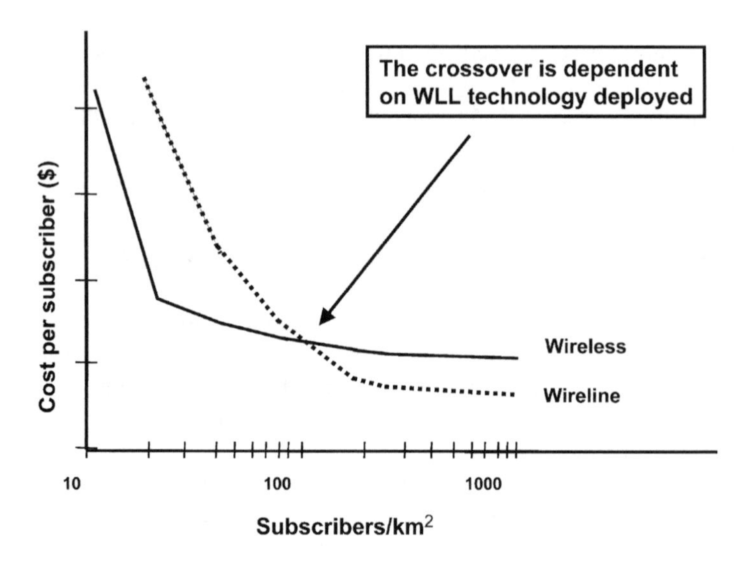

Figure 1.6. Comparison of capital costs between wireline and wireless access.

wireless (local loop) systems with respect to capital cost per subscriber as a function of subscriber density.

In a traditional copper-based external plant, a major part of capital expenditure is associated with the copper cable and wire, right-of-way acquisition, and installation of the vast branching network consisting of distribution cables, feeder plant, and drop wires into subscribers' premises. In contrast, most of the external plant for a WLL system consists of radio base stations and antennas whose major components are essentially *electronics*, whose costs generally decline rapidly over time. Furthermore, the installation of such equipment is much less labor intensive and time efficient, and therefore cost effective. The major cost component in WLL systems is the number of base stations that are required to provide the necessary coverage and quality of service.

As shown in Figure 1.6, the wireless access solution is cost effective (compared to wireline) when the subscriber densities (not population densities) are below about 200 subscribers/km^2. At higher subscriber densities (e.g., in urban areas), the number of base stations required may increase rapidly, thereby increasing the cost for acquiring base stations and the base station sites. However, it is important to note that the relationship shown and the crossover point shown in Figure 1.6 is dependent on the specific WLL technology. Consequently, this relationship will come into play when making a WLL technology selection for a specific deployment scenario.

Another positive attribute for wireless access that represents a cost advantage is illustrated in Figure 1.7. It shows the typical growth profiles for wireline and wireless external plants to meet the changing needs for subscriber growth. Wireline access infrastructure requires a large up-front investment followed by additional investments in relatively large steps to meet increasing subscriber growth.

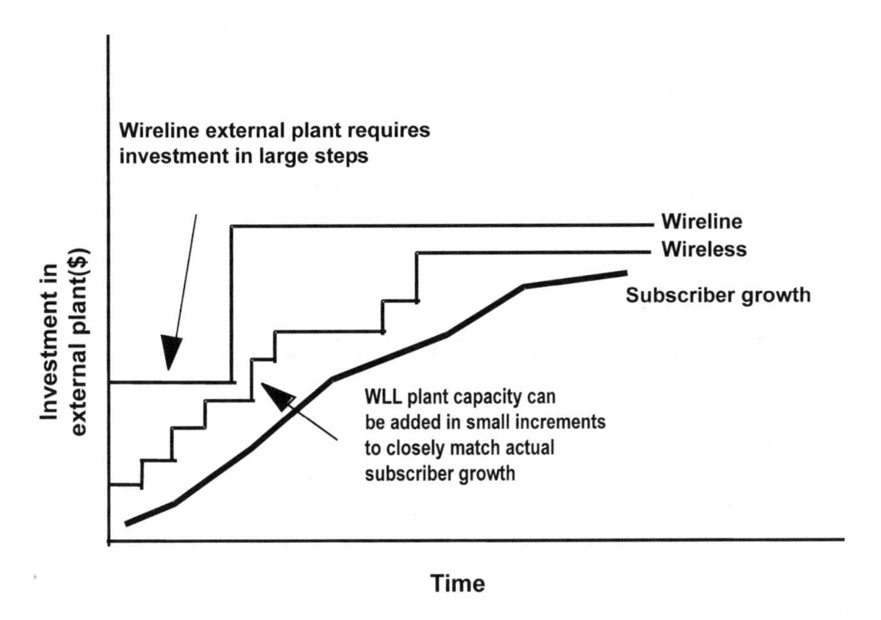

Figure 1.7. Comparison of growth profiles for wireline and wireless external plant.

A wireless access system, on the other hand, can start with a modest infrastructure (e.g., one or two base stations) to meet the initial demand, with the flexibility of adding more facilities in much smaller increments that can closely track the changes in subscriber growth. Sometimes this may only require the sectorization of antenna site at an existing base station or the addition of a cell base station to meet the incremental subscriber growth. Thus, the WLL operator can expect faster payback on investment and less financial exposure resulting from errors in subscriber growth forecasts.

Operational cost savings represent another advantage of employing wireless instead of wireline access solutions. In the case of copper-based wireline access networks, the trouble reports associated with the distribution cables, drop wires, and in-house wiring represent a significant component of a telephone company's maintenance activity and associated costs (around 35–40%).

In a wireless access network, these components are replaced by base stations and antennas which require significantly less number of trouble reports, dispatch activities, and repair activities, leading to a corresponding reduction in operations costs. Reduction in annual operating expenses in the range of 25% for each subscriber can be realized in a wireless access network. Furthermore, the nature and quantity of the WLL external plant makes it less vulnerable to theft and vandalism with associated reduction in maintenance costs. However, because the customer premises equipment (CPE) for WLL systems is more complex, potential for CPE maintenance problems may be greater than for fixed access.

Opportunity costs are associated with loss of revenue due to delays in provision of service to waiting subscribers. In case of WLL systems, the network implementation and service deployment times are comparable to those with cellular networks, which are much faster than fixed networks (months instead of years). Thus returns on investment for WLL systems can be achieved much faster, leading to reduction in so-called opportunity costs to the operator.

1.5 POTENTIAL LIMITATIONS OF WIRELESS ACCESS AND WLL SYSTEMS

As discussed in the previous section, there are many cost and operational advantages associated with WLL systems for providing telecommunication services. However, the following potential limitations of wireless access need to be recognized and factored in when choosing suitable access network technology:

- *Constraints on Spectrum Availability:* This is one of the major hurdles for deploying wireless access systems since the most appropriate frequency band may already be employed for alternative applications like cellular mobile or microwave systems and may not be available, thereby limiting the choice for a WLL system.
- *Available Quantity of Bandwidth:* Limitation on available bandwidth will impose a trade-off between number of customers that can be supported and the range of services and the quality of service which can be offered to customers.
- *Absence of Agreed International Standards:* The lack of agreed standards means that the manufacturers are offering WLL systems based on existing cellular mobile or cordless telecommunications technologies (for which the frequency spectrum may already be in use), or those based on proprietary technologies. In either case the operator is generally locked into dealing with a single equipment provider with little flexibility.
- *Need for Radio Link Planning:* Many WLL systems operate in frequency bands that require line-of-sight (LOS) transmission, so that for each customer equipment installation a LOS radio link (between the customer premises and the radio base station antennas) needs to be established to provide acceptable service quality.
- *Limitations on Quality of Service:* The voice quality and availability (in terms of assured bit error rates) for radio channels is subject to degradation caused by a number of factors like multipath fading, co-channel interference, and signal loss due to rain.
- *Concern for Security and Privacy:* This significantly limits the services that can be offered to customers requiring a high level of secure access and information flows, because radio transmissions are subject to intercept and

overhearing of conversations. However, the new digital systems can provide better protection and can be encrypted.

1.6 SOME DIFFERENCES BETWEEN WLL AND CELLULAR MOBILE SYSTEMS

Though there are some similarities between WLL and cellular mobile systems in terms of radio technologies and use of such components as base stations and use of cell structures for coverage, there are important differences in the planning and operational aspects of these systems. Some of the key differences are summarized in Table 1.2.

TABLE 1.2. Differences Between WLL and Cellular Mobile Systems

Item	WLL Systems	Cellular Systems
Frequency bands	No dedicated bands. Sharing with other applications may be needed.	Dedicated bands that are highly regulated.
Radio interfaces	Not standardized. Many proprietary interfaces.	Conform to regional and/or international standards.
Coverage	Noncontiguous. Coverage needed in areas where subscribers reside.	Contiguous. Coverage needed to support subscriber movements over wide areas.
Traffic load	Generally no spatial component. Subscribers not likely to move around over wide areas.	Need to take account of spatial traffic changes due to mobility.
Network performance	Set by wire-line performance expectations—more demanding.	Less demanding in order to factor in effects of mobility.
Performance seen by subscriber	Performance needs to be met for individual subscribers.	Performance set as an average over the coverage area.
Propagation conditions	Better conditions due to optimized location of directional antennas.	Omnidirectional antennas and in-building use adversely affects signal reception.
System roll-out	Gradual—according to increasing demand.	Significant initial roll-out to provide coverage for mobility.
Tariffs	Comparable to wire-line rates.	Higher tariffs to support mobility—terminating charges may apply.
Network management	Based on wire-line NM principles including the terminals.	Cellular terminals not covered by the NM system.

1.7 STANDARDIZATION STUDIES ON FIXED AND BROADBAND WIRELESS ACCESS SYSTEMS

Standardization of various aspects of fixed and broadband wireless access systems has been underway in international as well as regional standards development organizations (SDO) like the ITU and ETSI for some time. The scope of these studies include identification of suitable frequency spectrum, efficient methods for sharing of frequency spectrum between different wireless services, and specification of radio subsystems for efficient utilization of available frequency spectrum.

International Telecommunications Union (ITU) is one of the Specialized Agencies of the United Nations with the mandate to facilitate the development and implementation of telecommunications services worldwide. As shown in Figure 1.8, the activities of the ITU are partitioned into three broad sectors with broad areas of activities assigned to *Study Groups (SG)* in each sector.

The three ITU sectors have the following broad mandates:

- ITU-R: Radio communications-related standardization, along with regulation and management of radio-frequency spectrum and satellite orbits.
- ITU-T: Setting of telecommunication standards in order to facilitate interconnection of worldwide telecommunication networks and delivery of telecommunication services.

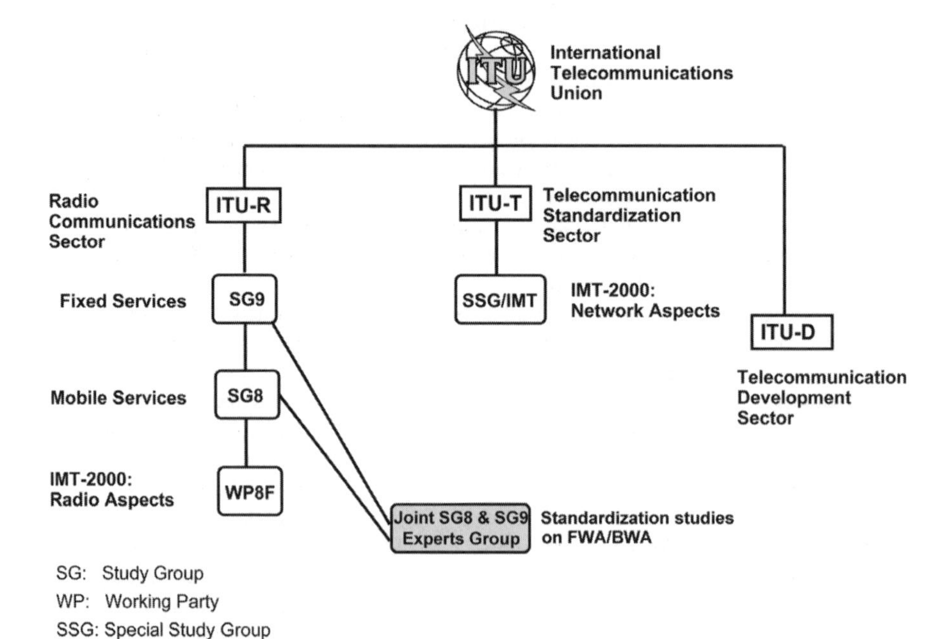

Figure 1.8. ITU structure and study groups involved in WLL-related studies.

- ITU-D: Providing technical assistance and promoting international cooperation for the improvement of telecommunications infrastructure in developing countries.

ITU-R (International Telecommunications Union—Radio Communications Sector) has been playing a leading role in the study, standardization, and radio spectrum aspects of wireless access with a definite slant towards applications of wireless access in developing countries. As mentioned in Section 1.1, ITU-R Recommendations and documents related to the applications of wireless access for providing services to fixed (as opposed to mobile) terminals/subscribers use the term FWA instead of WLL.

As shown in Figure 1.8, ITU-R has established a Joint Group with experts from Study Group 8 (ITU-R SG8) and Study Group 9 (ITU-R SG9) which is conducting extensive studies that address the frequency spectrum requirements, frequency sharing plans, and performance and quality of service aspects of wireless access systems for fixed and broadband services. Besides publishing a *Handbook on Fixed Wireless Access*, which is a good reference document, a large number of relevant Recommendations have also been developed by ITU-R.

Besides the ITU-R SG8 and SG9 Joint Experts Group, some studies related to the convergence of fixed and wireless services and networks are also being carried out in ITU-R WP8F (IMT2000: Radio Aspects) and ITU-T SSG/IMT (IMT2000: Network Aspects). The objective of these studies is to develop access and core network architectures for third-generation (3G) mobile communication systems that will ensure seamless and transparent delivery of services to end users irrespective of the access method being used.

European Telecommunications Standards Institute (ETSI) has also been conducting extensive studies on various aspects of fixed and broadband wireless access. ETSI has established a separate project called broadband radio access network (BRAN) for specification of fixed and cordless broadband access for data rates from 25 Mb/s to 155 Mb/s and coverage ranging from 50 m to 5 km. Standards like HIPERLAN, HIPERACCESS, and HIPERLINK fall under the BRAN mandate. Prior to establishment of the BRAN project, ETSI had developed numerous specifications and technical reports related to radio in the local loop (RLL)—especially directed to the use of DECT and GSM radio for RLL applications.

IEEE in the United States has also been developing specifications for broadband wireless access systems for different coverage ranges and different frequency bands as part of its activities on wireless local area networks (WLANs). These specifications include 802.11, 802.11a, 802.16, and 802.16a.

The above standardization activities in ITU-R, ETSI-BRAN, and IEEE 802 committees are being conducted in cooperation and consultation with each other, as well as in cooperation with other relevant organizations like IETF (Internet Engineering Task Force) and ATMF (Asynchronous Transfer Mode Forum). Many of the standards documents from these standards development organizations that have direct relevance to the contents of the book are included in the Bibliography.

CHAPTER 2

Fundamentals of Radio Systems

WLL systems are a class of radio systems and utilize a part of the radio-frequency spectrum for establishing point-to-multipoint communication between a base station and the remote subscriber terminals. Radio-frequency spectrum is a common, limited resource that needs to be shared, managed, and utilized in an efficient and effective manner. In order to provide continuous two-way radio communication for multiple subscribers, the frequency spectrum made available by the regulatory authority needs to be partitioned into suitable radio channels using appropriate duplexing and multiple access techniques. Furthermore, the radio transmitter system needs to provide such basic functions like source coding, channel coding, interleaving, and modulation, with appropriate complementary functions provided at the radio receiver. This chapter is intended to provide an overview of basic radio system requirements and functions that will form the basis for more detailed radio aspects specific to WLL systems discussed in latter chapters.

2.1 RADIO SPECTRUM AND FREQUENCY BANDS

The total range of frequencies at which radio communication is possible is called the *radio spectrum*, which spans from about 100 kHz to 300 GHz. As shown in Table 2.1, the range of frequencies that are identified for wireless access applications can range from 400 MHz to 40 GHz. This range of frequencies is divided into "frequency bands," which are used for different wireless access applications like cellular mobile, low-power cordless telecommunications, wireless local loop (WLL), wireless local area networks (WLANs), multipoint distribution systems (MDSs), etc. Though a frequency band may be identified by a single frequency (like 900-MHz band), the width of the frequency spectrum allocated to the band may vary from system to system within the same application. For example, the 900-MHz band for GSM covers an allocation of 25-MHz duplex paired frequencies between 890–915 and 935–960 MHz, whereas for the Japanese PDC digital cellular system the 900-MHz band covers frequencies between 810–826 and 940–956 MHz.

Introduction to WLLs. By Raj Pandya
ISBN 0-471-45132-0

TABLE 2.1. Radio-Frequency Bands for Wireless Access Applications

Frequency Band	Application
400 MHz	Analog cellular systems: NMT and TACS
800 MHz	Analog cellular systems: AMPS, NMT
	Digital cellular systems: IS-136 (TDMA), IS-95 (CDMA)
900 MHz	Digital cellular systems: GSM (TDMA), PDC (TDMA)
1.4 GHz	Subscriber radio: Point-to-multipoint
	Digital cellular systems: PDC
1.8 GHz	Digital cellular: GSM-1800 (DCS)
	Digital cordless: DECT, PHS
1.9 GHz	Personal Communication Systems in North America: frequency up-shifted versions of IS-136, IS-95, GSM, etc.
2.4–2.6 GHz	Point-to-multipoint and multipoint distribution systems; wireless local area networks (IEEE 802.11)
3.4–3.6 GHz	Some proprietary WLL) systems; also proposed for WLL systems by ITU-R
5.0 GHz	Wireless local area networks (HIPERLAN)
10.5 GHz	Point-to-multipoint systems
Various bands between 11 and 40 GHz	Various broadband wireless access (BWA) systems like LMDS, MVDS, IEEE 802.16, and ETSI-BRAN HIPERACCESS

In order to ensure that the radio spectrum, which is a collective and very critical global resource, is used most efficiently and with minimum interference, the ITU has been assigned the responsibility for managing the radio spectrum on a global basis. The Radio Communication Sector of the ITU (ITU-R) sets out in a series of *Recommendations* and *Reports* the radio planning and operating procedures to be followed internationally. ITU-R administers the specification of technical radio standards, the registration of radio users, and the allocation of radio bandwidth to ensure that interference of signals is minimized.

ITU-R carries out this function in coordination with two other ITU organizations: the IFRB (International Frequency Registration Board) and the WRC (World Radiocommunication Conference). Frequency bands are identified internationally by ITU-R for such applications as mobile communication, maritime communication, terrestrial fixed network communication, satellite communication, and so on. However, the actual use of the radio spectrum, the licensing of its use, and the detailed registration of the authorized users is the responsibility of the individual countries and their radio regulatory bodies (like the Federal Communications Commission in the United States).

The ITU frequency spectrum plan divides the world into three major regions: Region 1, which essentially covers Europe and Africa; Region 2, which covers North and South America; and Region 3, which covers the Asia Pacific region. The two regional bodies most influential in setting the regional plans include the FCC

(Federal Communications Commission) in the United States and the CEPT (Conference of European Post and Telecommunications) in Europe. Based on the frequency band allocation by ITU-R, the regional bodies further split the spectrum into a larger number of smaller bands, each allocated for specific applications like public land mobile networks (PLMNs), microwave radio, satellite uplink, and downlink.

Unfortunately, the regional and national regulatory bodies historically have not coordinated their frequency band allocations to similar services leading to some obvious problems in the operation and future evolution of mobile communication services. For example, the 900-MHz band used for GSM throughout Europe as well as in many other Asian countries is allocated by FCC for use in the unlicensed band for cordless telephones, thereby making roaming of European (900-MHz) GSM phones in United States impossible.

2.2 CLASSIFICATION OF RADIO SYSTEMS

Radio communication systems in different forms and for different applications have been in existence for over a century. Radio systems can be broadly classified as point-to-point (PTP), point-to-multipoint (PTM), and any-point-to-any-point (APAP) systems. Point-to-point microwave radio systems are commonly used in public telecommunication networks in the core networks as well as for backhauling traffic from concentrators, distribution points, and base stations in the access networks to the PSTN. These point-to-point microwave systems operate in the GHz bands, using line-of-sight paths and deploy highly directional antennas. Figure 2.1 illustrates the use of a PTP radio system to backhaul traffic from a WLL base station to the local switch, and it also illustrates the use of a PMP radio system to provide wireless access to the end users by the radio base station.

The PTM system illustrated in Figure 2.1 comprises a base station serving a number of remote stations. Cellular and WLL systems fall under this category. PTM systems are considered asymmetric in the sense that the down-link (base-to-remote terminal) and the up-link (remote terminal-to base) need to be considered separately

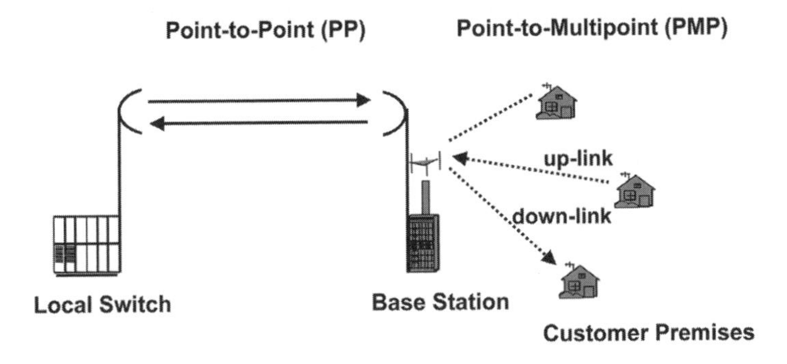

Figure 2.1. Illustration of a point-to-point and a point-to-multipoint radio systems.

and differently. In the down-link direction the base station needs to broadcast the signal to all the remote terminals within its serving cell or sector using an omnidirectional (or suitable sectored) antenna. In the up-link direction, some form of "multiple access" method needs to be employed so that the transmission from the remote terminal reaches the base station without interference from the transmissions from other remote terminals being served by the same base station. If the remote terminal is in a fixed location in a WLL system, it can use a directional antenna to achieve better communication.

Any point-to-any-point radio systems are characterized by the fact that all radio stations can be in communications with all other radio stations in the system at any given time. Wireless local area networks (WLANs), like IEEE 802.11operating in the 2.4-GHz band and ETSI HIPERLAN operating in the 5-GHz band, are examples of APAP radio systems. These generally operate in unlicensed radio frequency bands, and their range is restricted to a few hundred meters.

2.3 FULL DUPLEX RADIO COMMUNICATION AND DUPLEXING METHODS

Full duplex communication is required in order to support continuous and simultaneous communications in both directions. As shown in Figure 2.2, it requires both a transmitter and a receiver at each of the two ends of the communications link (i.e., at the subscriber premises and the radio base station).

There are two basic methods to achieve full duplex communications: frequency division duplex (FDD) and time division duplex (TDD).The principles for FDD and TDD operation are illustrated in Figure 2.3.

In the case of FDD operation the up-link and down-link signals are separated in frequency, and two distinct blocks of frequencies are assigned for this purpose. In the time domain, the up-link and down-link signals traverse the radio link at the same time. FDD operation is characterized by the following properties:

- Suitable for large distances/cells used in cellular or WLL systems
- Suitable for high-power applications like cellular mobile

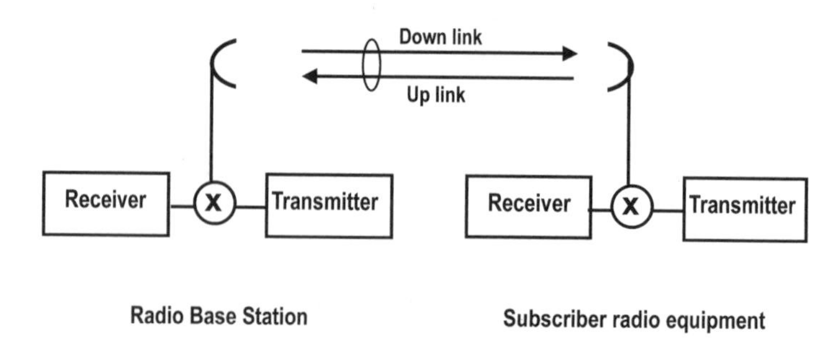

Figure 2.2. Full duplex communication for point-to-multipoint radio systems.

Frequency Division Duplex (FDD)

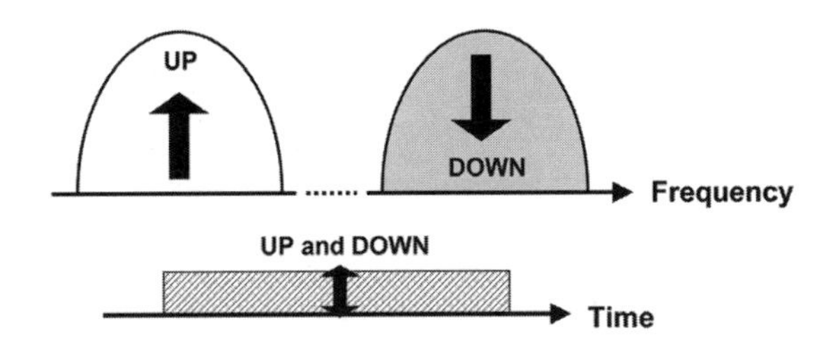

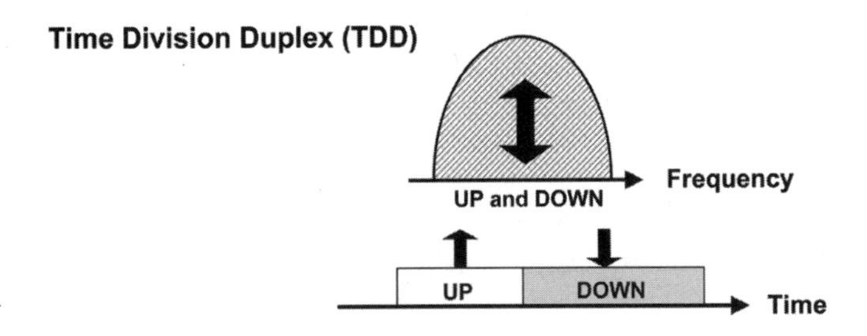

Figure 2.3. Principles of frequency division and time division duplex operation.

- Does not require frame synchronization in the time domain
- Requires two distinct blocks of frequencies with adequate separation (guard band)

In the case of TDD operation the up-link and down-link signals are separated in time by assigning a set of distinct time periods (slots) for up-link and down-link signals. A single block of frequencies is used in the TDD operation. TDD operation has the following properties:

- Suitable for short distances (indoor and campus environments)
- Suitable for low-power applications like cordless telecommunication systems
- Suitable for asymmetric operation where different amount of bandwidth (time slots) can be assigned for up-link and down-link directions
- Requires frame synchronization for proper operation
- Requires a single block of frequency spectrum

The two radio-frequency bands assigned for FDD or the single radio-frequency band allocated for TDD are generally divided into a number of smaller and generally equal units referred as radio *channels*. Each radio channel is specified by its center frequency, and the bandwidth is assigned to it. Consecutive channels are separated by fixed guard bands to minimize overlap. A channel *raster* is used to quantify the constant bandwidth allocated to individual channels and the guard band between channels.

In the case of FDD, the available spectrum has two components (which may or may not be contiguous) and the allocated spectrum is referred to as *paired* spectrum. The lower segment (i.e., lower set of frequencies) is generally used for up-link (terminal-to-base station) transmissions, and the upper segment (higher set of frequencies) is used for down-link (base station-to-terminal) transmissions.

As an example of FDD duplex operation, Figure 2.4 shows the FDD channel structure for a GSM cellular mobile system where the frequency spectrum for GSM consists of the following two segments:

890–915 MHz: Up-link (mobile-to-base station b).
935–960 MHz: Down-link (base station-to-mobile), with a separation (guard band) of 20 MHz between the two blocks.

The GSM frequency bands are then divided into 124 full duplex channels or carriers of 200-kHz width. Thus, the (channel) spacing between the up-link frequencies and down-link frequencies in each duplex channel is 45 MHz. The duplex channel spacing is chosen to maximize the usage of the available spectrum within the allocated band, but at the same time to minimize the possibility of radio interference between transmit and receive directions. In GSM, each duplex channel (carrier) is divided into eight time slots for time division multiple access (TDMA) operation.

As opposed to FDD, where one-half of the frequency spectrum is allocated for up-link transmission and the other half for down-link transmission, in case of TDD

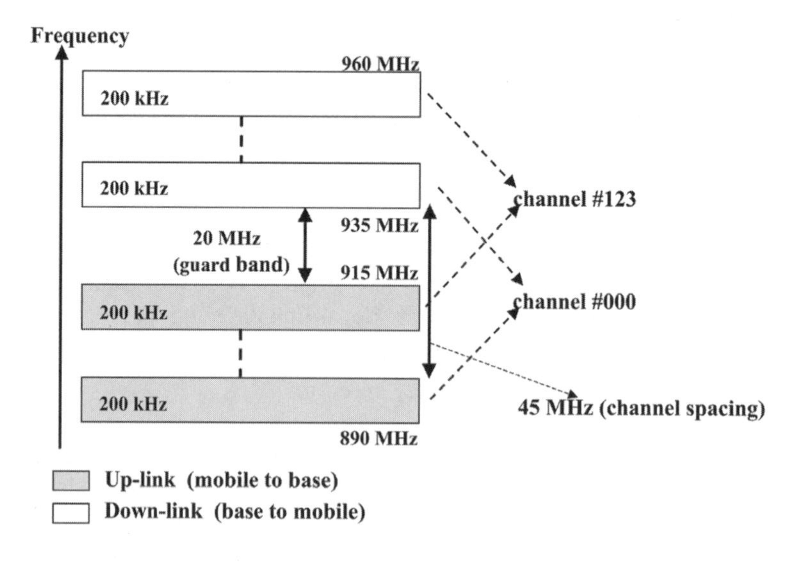

Figure 2.4. Example of an FDD channel structure used for a GSM system.

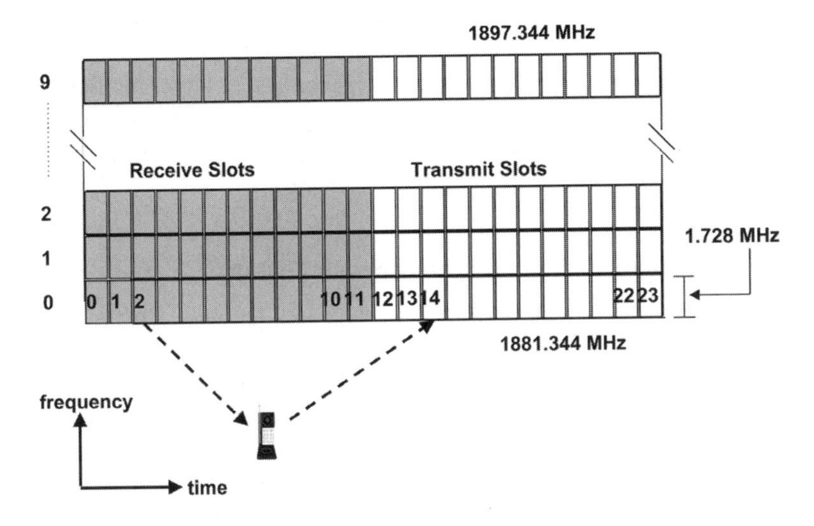

Figure 2.5. Example of a TDD channel structure used for a DECT system.

there is no pairing of the spectrum and the entire allocated frequency spectrum can be used for up-link as well as the down-link transmissions. The separation of the signals in the two directions is achieved by dividing the frequency spectrum into time slots and then pairing the time slots (rather than the frequencies as in FDD) for full duplex operation. DECT cordless telecommunication system is an example of TDD mode of full duplex operation, and its channel structure is illustrated in Figure 2.5.

In the DECT system the 20 MHz of spectrum (1880–1900 MHz) is divided into 10 frequency channels (1.728 MHz + guard band). Each frequency channel is then divided into 24 time slots providing 12 full duplex channels (*paired* slots) per carrier, leading to a total of 120 potential simultaneous connections. Thus a full duplex channel in DECT is identified in terms of the frequency channel + the pair of time slots in that frequency channel.

An advantage of TDD is its ability to dynamically adjust the relative amounts of time slots for receive and transmit directions, thereby providing asymmetrical connections between the host and the terminal for data communications.

2.4 MULTIPLE ACCESS METHODS FOR WIRELESS ACCESS SYSTEMS

Generally a fixed amount of frequency spectrum is allocated for a specific application like cellular mobile or WLL service by the national regulator (e.g., the FCC in the United States). Multiple access techniques are then deployed so that many users can share the available spectrum in an efficient manner. Multiple access systems specify how signals from different sources can be combined efficiently for transmission over a given radio-frequency band and then separated at the destination, minimizing mutual interference. The three basic multiple access methods currently in use in radio systems are: frequency division multiple access (FDMA), time division multiple access (TDMA), and code division multiple access (CDMA).

2.4.1 Frequency Division Multiple Access (FDMA)

The radio channel structure for frequency division multiple access (FDMA) systems is illustrated in Figure 2.6.

In the case of FDMA, users share the available spectrum in the frequency domain and a user is allocated a part of the frequency band called the traffic channel. The user's signal power is therefore concentrated in this relatively narrow band in the frequency domain, and different users are assigned different traffic (frequency) channels on a demand basis. Interference from adjacent channels is minimized by the use of guard bands and band-pass filters that maintain separation of signals associated with different users. In FDMA systems that utilize the FDD duplex method, a radio traffic channel is defined by a pair of frequency channels—one for the up-link transmissions (subscriber-to-base station) and the other for down-link transmissions (base station-to-subscriber). Analog cellular systems like AMPS, TACS, and NMT are examples of FDMA/FDD applications.

The principle of operation of FDMA/FDD radio systems is illustrated in Figure 2.7, where, during the radio resource allocation phase of the call setup, the four subscribers are assigned transmit (up-link) frequency channels f1, f2, f3, and f4 and receive (down-link) frequency channels f1*, f2*, f3*, and f4*. The subscriber terminals therefore transmit on these assigned frequencies. When these signals

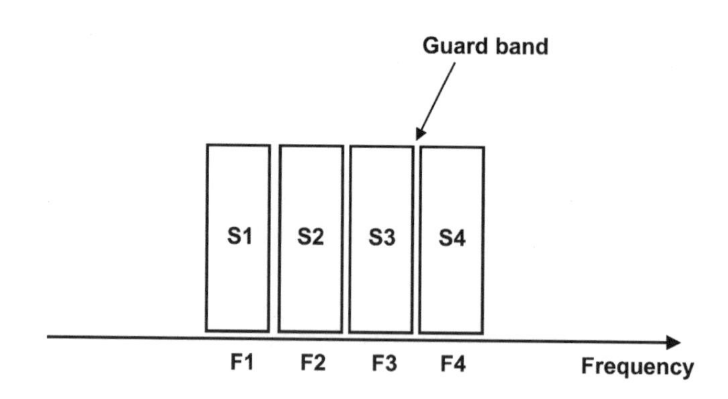

Figure 2.6. Radio channel structure for FDMA systems.

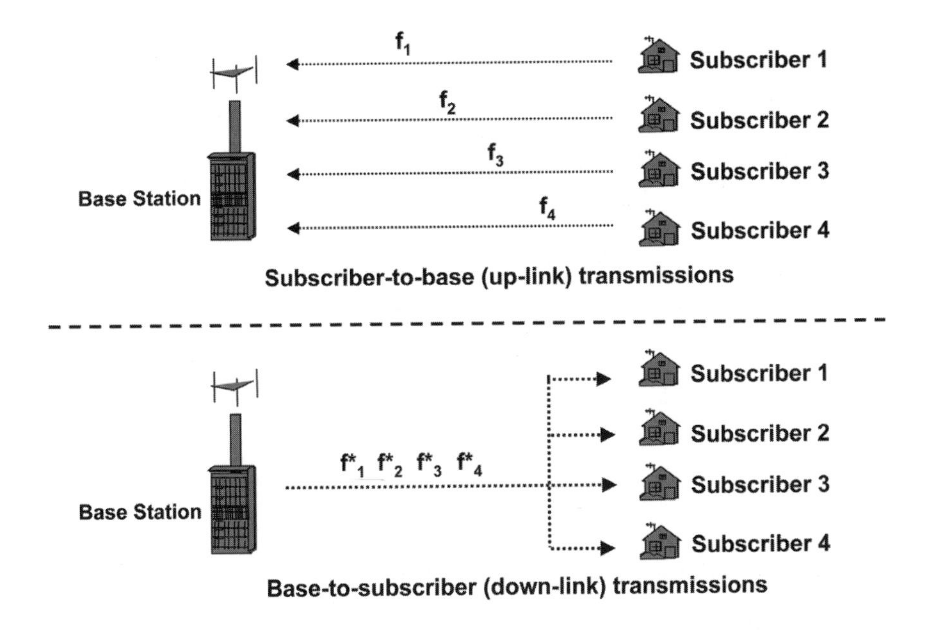

Figure 2.7. Principle of operation for FDMA/FDD systems.

arrive at the base station, they are separated using band-pass filters for further processing.

At the base station, the information intended for the individual subscribers is used to modulate the assigned FDD paired frequencies (f1*, f2*, f3*, and f4*) which are then passed through a *combining* function, and the resulting signal is transmitted. The signal intended for an individual subscriber is extracted by the subscriber terminal using a filter tuned to its assigned *receive* frequency (i.e., f1*, f2*, f3*, or f4*).

2.4.2 Time Division Multiple Access (TDMA)

In the TDMA techniques that are utilized in many digital cellular systems, the available spectrum is partitioned into narrow frequency bands or frequency channels (similar to FDMA), which in turn are divided into a number of time slots. The radio channel structure for a TDMA system is illustrated in Figure 2.8. An individual user is assigned a time slot that permits access to the frequency channel for the duration of the time slot. Thus, the traffic channel in case of TDMA consists of a time slot in a periodic train of time slots that make up a frame. For example, each 200 kHz frequency channel in GSM (see Figure 2.3) is divided into 8 time slots (full rate). In case of TDMA systems, guard bands are needed both between frequency channels and between time slots.

The principle of operation of TDMA/FDD radio systems is illustrated in Figure 2.9, where, during the radio resource allocation phase of the call set up, the four subscribers may be assigned a transmit (up-link) frequency channels fn and receive (down-link) frequency channels fn* and individual time slots T1, T2, T3,

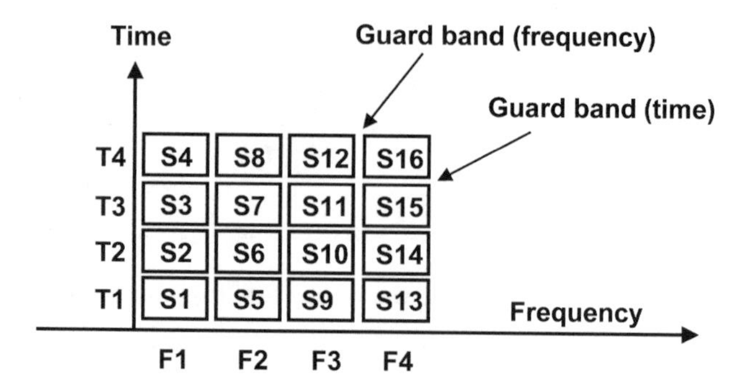

Figure 2.8. Radio channel structure for TDMA systems.

and T4 on these up-link and down-link frequency channels. The subscriber terminals therefore transmit on these assigned time slots in frequency channel fn. When the signals arrive at the base station, information on individual time slots is extracted for further processing.

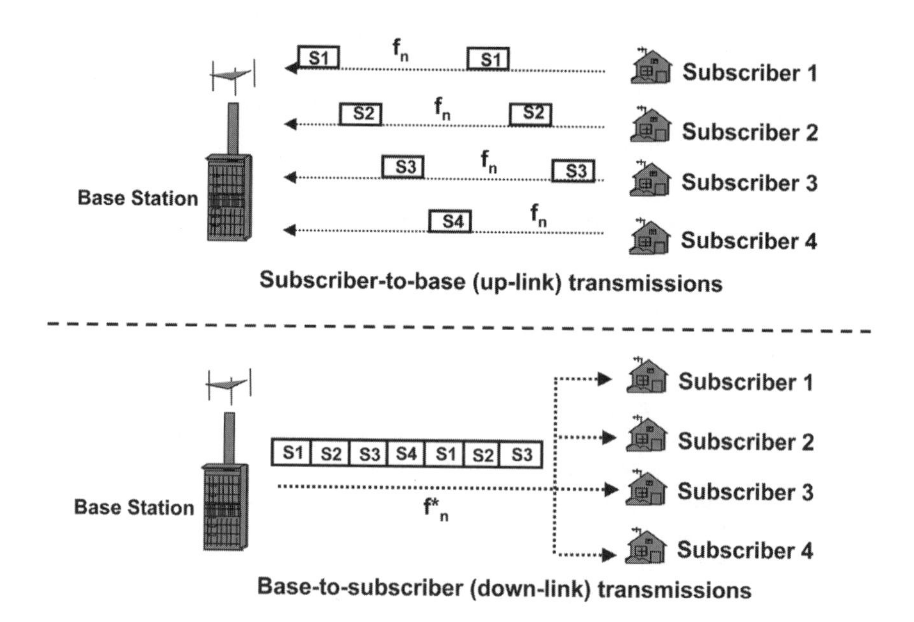

Figure 2.9. Principle of operation for TDMA/FDD systems.

At the base station, the information intended for the individual subscribers is placed on the assigned time slots, which are transmitted using the paired FDD frequency channel fn*. The subscriber terminals then extract the information from their assigned time slots.

Thus, in the above example TDMA system, the entire frequency carrier is used by each of the four subscribers in turn, each using the frequency carrier for one-quarter of the time. A burst of information is sent by each subscriber terminal in a predetermined time slot allocated by the base station. The base station, however, transmits continuously on its transmit frequency, with each subscriber terminal selecting the information only from the time slot that is allocated to it.

2.4.3 Code Division Multiple Access (CDMA)

CDMA systems utilize the spread spectrum technique, whereby a spreading code [generally a pseudorandom noise (PN) code] is used to allow multiple users to share a block of frequency spectrum. In CDMA systems like the IS-95 CDMA cellular system the (digital) information from an individual user is modulated using a unique spreading sequence assigned to that user, and the resulting signals from different users are then transmitted over the entire CDMA frequency channel (e.g., 1.25 MHz in case of IS-95). At the receiving end, the desired signal is recovered by using a de-spreading process. Since the signals in the case of CDMA utilize the entire allocated block of spectrum, no guard bands of any kind are necessary within the allocated block. The radio channel structure for CDMA systems is illustrated in Figure 2.10.

The principle of operation of CDMA/FDD systems is illustrated in Figure 2.11, where the information from individual subscribers is spread using unique spreading

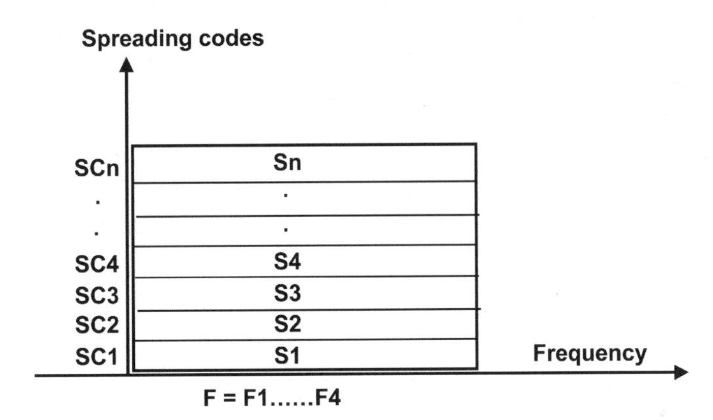

Radio channel Id = PN Code Id

Fx = Frequency channel x
SCx = Spreading Code x
Sx = Subscriber x

Figure 2.10. Radio channel structure for CDMA systems.

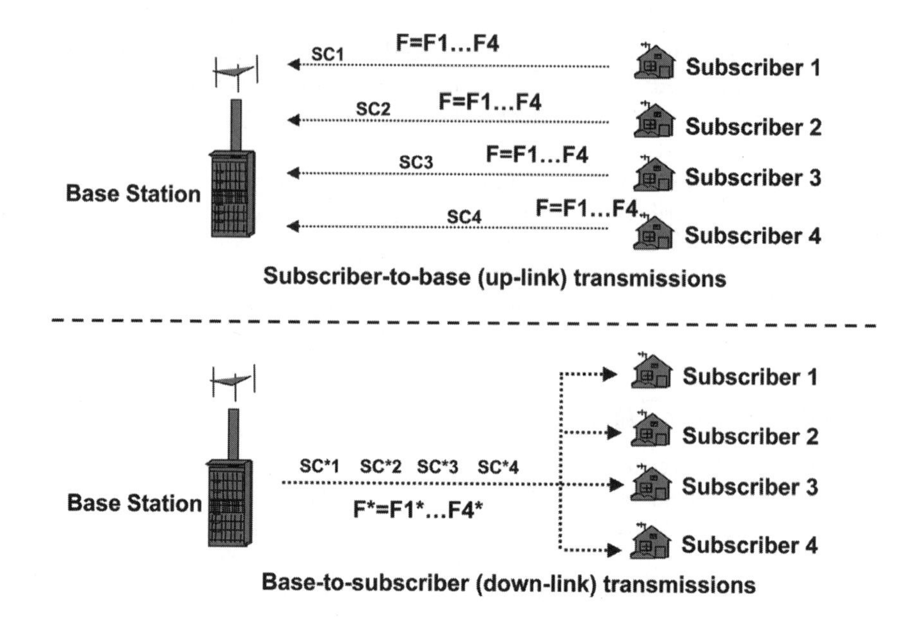

Figure 2.11. Principle of operation for TDMA/FDD systems.

codes (SC1, SC2, SC3, etc.) assigned to them at radio resource allocation phase of the call setup. The resulting transmit signals from individual subscribers occupy the entire up-link frequency band (F) allocated for CDMA operation. At the base station, individual subscriber information is extracted by de-spreading it with a copy of the spreading sequence for the individual user in the receiving correlator.

In the down-link direction, the information for individual subscribers is spread over the allocated down-link frequency band (F*) using a different set of spreading codes (SC*1, SC*2, SC*3, etc.) and are combined for transmission. Again, the individual subscriber terminals extract their signals using a de-spreading process on the combined signal received from the base station.

Cellular and cordless systems based on TDMA and CDMA principles, which are generally the basis for design and operation of many WLL systems, are discussed in greater detail in Chapter 3.

2.5 MAJOR FUNCTIONS FOR A RADIO SYSTEM

Figure 2.12 illustrates the key functions associated with a simple radio system consisting of a transmitter and a receiver. The analog information generated by the source (voice, voice-band data, or video) is first digitally encoded by the source coder. The function of the channel encoder that follows is to encode the digital signal for error detection and/or error correction. The interleaving process is employed to minimize the effects of block errors, which generally cannot be

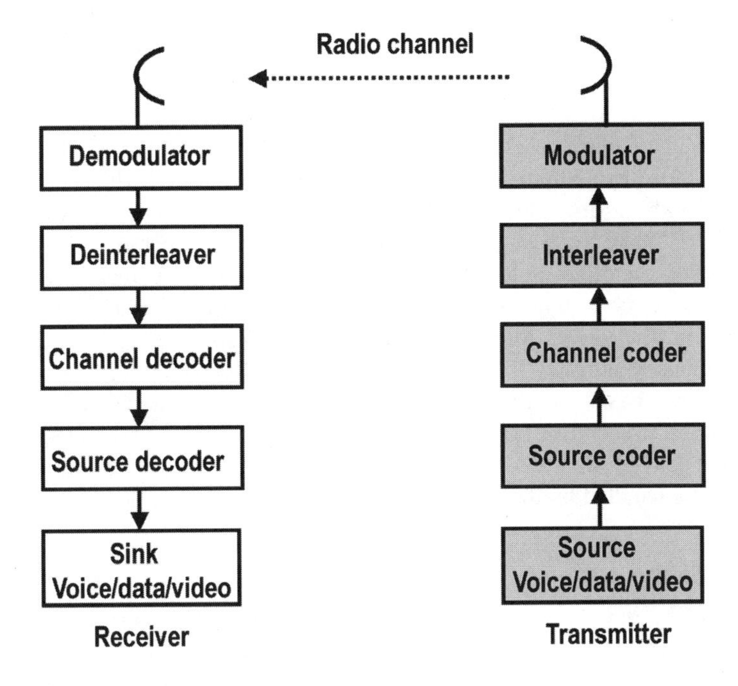

Figure 2.12. Major functions associated with a simple radio system.

mitigated by the channel encoder. The function of the modulator is to up-shift the signal to a suitable higher frequency so that it can be transmitted over the radio path in the most efficient manner (i.e., maximize spectrum efficiency). At the receiver, a reverse process is deployed to extract the original signal. Some of the functions of the radio systems that are important for efficient operation of a WLL system are briefly described below.

2.5.1 Source Coding

Source coding involves conversion of analog (speech) signals into a digital signal. Ordinarily, an A/D converter may be used which generates a 64-kb/s PCM (pulse code modulation) signal. The PCM is based on the assumption that the highest audible frequency contained in a voice signal is 4 kHz, so that using an 8-kHz sampling rate (Nyquist rate) and an 8-bit/sample representation results in a 64-kb/s bit rate. However, from an information-theoretic perspective, the 64-kb/s PCM source coding is very inefficient because it does not take into account the predictability of speech signals. ADPCM (adaptive differential pulse code modulation), which quantifies the difference between the signal level at the previous sample and the current level (instead of the absolute signal level used in ordinary PCM), results in a 32-kb/s rate with very little degradation in voice quality compared to the 64-kb/s PCM signal.

In order to maximize spectral efficiency, more efficient voice encoding methods are deployed in cellular systems. These voice coders (or vocoders) model the vocal tract of the speaker based on the first few syllables and then send the information on how the vocal tract is being excited. For example, the GSM utilizes a coder known as a residual pulse excitation–long-term prediction (RPE–LTP) which produces an output of 13 kb/s. More efficient voice coding methods like enhanced variable rate coding (EVRC) are increasingly being deployed in order to conserve bandwidth.

Obviously, there is a trade-off between the voice quality perceived by the end user and the speech coding efficiency. If wire-line voice quality is a requirement in a WLL system, then ADPCM is probably the best option. Vocoders produce digital speech signals at bit rates that are much lower than the 64-kb/s PCM rate that is currently the basis of digital telecommunication networks (switching and transmission). Generally, transcoding and rate adaption (TRAU) functionality is required to convert the vocoder output into a standard 64-kb/s PCM stream that can efficiently be handled by the network.

2.5.2 Channel Coding

The purpose of channel coding is to mitigate the effects of interference and fading generally encountered on radio channels. Transmission over radio links is subject to significantly greater interference and delay than fixed network links. Raw bit error rates in the range of one per thousand are not uncommon on radio links, so that error correction protocols are essential for data transmission. Forward error correction methods, which are designed to ensure bit error rates in the range of one per million which are encountered in fixed network links, may lead to unacceptable residual errors. Error correction methods attempt to correct errors introduced by the radio channel by adding redundancy to the transmitted signal. Broad classes of channel coding schemes used in mobile communication systems include *block coding*, *convolution coding*, *turbo coding*, and *adaptive multirate (AMR) coding*.

In the case of block coding, the information to be transmitted is arranged in a matrix that is then multiplied by another matrix (block code), resulting in a *code word* that is also transmitted. The contents of the block code (matrix) are fixed and depend on the coding scheme used. The block code is known to both the transmitter and the receiver. The receiver again arranges the received signals into an identical matrix and multiplies it by the block code. If the resulting *code word* at the receiver does not match the one received from the transmitter, the received information is in error. The error correction then requires further (complex and processing intensive) matrix manipulations. Well-known block codes that are frequently deployed include BCH (Bose–Chaudhri–Hocquenghem) and R–S (Reed–Solomon) codes.

In the case of convolution coding, redundancy is added to the signal by combining the incoming bits with a suitable mask. The decoder in the receiver can use the knowledge of the introduced redundancy to correct the errors. The most commonly used decoder used in mobile systems is the so-called Viterbi decoder. Convolution coding is considered as a more powerful error correction scheme than block coding and requires less processing. However, it cannot be used for error detection and therefore cannot support ARQ (automatic repeat request) operation for mitigating effects of channel interruptions or handoffs.

Turbo codes, first introduced in 1993, are essentially parallel-concatenated convolutional codes. The concatenated codes are separated by an interleaving mechanism. The turbo codes can provide exceptional error correction performance and are increasingly being deployed in third-generation (3G) mobile systems. However, they require increased levels of computing power and complexity for decoding.

The adaptive multirate coding (AMR) concept allows a flexible dynamic adaptation of the bit rate between source coding and channel coding, depending on the observed channel quality. The overall speech quality is improved by increasing the amount of error protection while reducing the bit rate for source coding if the channel quality worsens. When applied to a mobile system like GSM, the rate adaption is controlled by a dedicated in-band signaling procedure.

Forward error correction (FEC) and interleaving are used to take care of such impairments as channel fading and random noise or interference on the radio link. However, FEC will not be able to counter the effects of channel interruptions (and handoffs in cellular systems). Thus, depending on the required error performance, additional ARQ (automatic repeat request) protocols may be required where the receiver detects errors in a block of received data and requests retransmission of the block if an error is detected. It should be noted that convolution coding does not support error detection and therefore cannot be used to implement an ARQ protocol.

2.5.3 Interleaving

The main purpose of interleaving is to protect the transmitted signal against block errors that cannot be efficiently corrected by channel coding schemes—which work best against random errors. In fading situations, errors tend to occur in blocks, with the length of the block of errors depending on the depth of the encountered fading. The interleaving process essentially randomizes block errors so that the error correction algorithms can operate on them successfully.

An interleaver is a simple device that places the input bits in a matrix, filling it from left to right (i.e., row-wise). When the matrix is full, the data are read out in columns and transmitted. At the receiver, the data fill up the matrix from top to bottom in columns and, when full, reads them out in rows from left to right. The operation of an interleaver is illustrated in Figure 2.13.

Interleaving has one disadvantage in that it introduces delay while the matrix at the transmitter is filled and while the matrix at the receiver is filled. This delay is undesirable for speech communication, and the system designer needs to balance the reduction in error (hence improvement in speech quality) with the undesirable effect of interleaving delay.

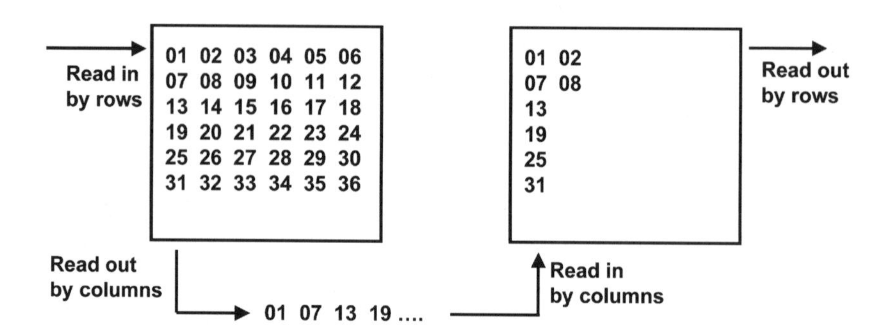

Figure 2.13. Example of implementing an interleaving function in a radio system.

2.5.4 Modulation

In a digital radio system, there is a speech-encoded and channel-encoded digital data stream consisting of square-wave binary pulses. This is the digital signal that needs to be modulated for efficient transmission over the radio channel. Amplitude modulation (AM) and phase modulation (PM) are the two basic classes of modulation methods. Frequency modulation (FM) is considered a subclass of phase modulation (also referred to as quadrature modulation), and in this case it is the frequency of the transmitted carrier that is changed in accordance with the variations in the base-band signal.

Broadly speaking, amplitude modulation changes the amplitude of the transmitted carrier according to the user-generated information signal that can be extracted at the receiving end by the reverse process of demodulation. Phase modulation, on the other hand, changes the phase of the transmitted carrier according to the variations in the modulating signal. Typically, PM is less susceptible to interference in the radio path. It is possible to combine AM and PM, which leads to QAM (quadrature amplitude modulation). Modulation methods are described in greater detail in the next section.

2.6 BASIC AND HIGHER-LEVEL MODULATION METHODS

2.6.1 Amplitude Modulation (AM)

Modems employing amplitude modulation alter the amplitude of the carrier signal, so that the carrier signal amplitude matches the original analog signal. For AM to work properly, the frequency of the carrier must be much higher than the highest frequency in the base-band signal. This ensures that the carrier signal can faithfully track and record even the fastest amplitude changes in the modulating analog signal. Amplitude modulation is simply carried out by using the modulating signal to control the power output of a carrier signal amplifier. The amplitude demodulation involves filtering out the carrier using an appropriate filter.

In the digital version of amplitude modulation, the amplitude of the carrier is varied between suitable finite amplitude and zero amplitude. These two amplitude states correspond effectively to the *on* and *off*, respectively, to values "1" and "0" of the modulating bit stream. Alternatively, two different amplitude values of the carrier signal may be used to represent "1" and "0."

2.6.2 Frequency Modulation (FM)

In the case of frequency modulation (FM), it is the frequency of the carrier signal that is altered to carry the signal content of the modulating digital bit stream, the amplitude, and the phase of the carrier remaining unchanged. Frequency modulation is achieved by simply mixing the modulating and carrier signals. The interference of the two signals during mixing leads to intermodulation creating side bands near the carrier frequency.

A frequency f in the original signal creates intermodulation products at frequencies $fc - f$ and $fc + f$, where fc is the carrier frequency. An original modulating signal occupying frequencies between fa and fb is thus mapped into two side bands. The lower side band extending from $fc - fb$ to $fc - fa$, and the upper side band from $fc + fa$ to $fc + fb$. The side bands thus mirror the original base band.

It is normal to filter the signal after modulation to remove one of the side bands (single side-band mode), and in some cases the carrier is also removed (suppressed carrier mode). This signal is then amplified for radio transmission. The benefit of the suppressed carrier mode is that the transmitter power is not wasted. At the receiver, demodulation of an FM signal is achieved by once again mixing the carrier frequency with the received side-band signal. One of the side bands produced by this process is the original signal, which can be filtered out. A disadvantage of the suppressed carrier operation is that an accurate oscillator and signal generator is needed at the receiver, thus adding to the receiver cost.

Generally, on–off keying (OOK) or frequency shift keying (FSK) are used for frequency modulation of digital signals. An OOK radio simply modulates the radio carrier by switching it on or off to represent "1" or "0" value for the digital signal.

FSK is a slight advance on OOK, in which the radio carrier signals of two or more different frequencies (both located within the bandwidth of the radio channel) are used to represent different "1" or "0" bit values of a digital bit stream. The simplest form of FSK is 2FSK in which two different frequencies are used to represent the binary bit values. The advantage of FSK over OOK is that the incoming signal is always present so that the receiver can always tell that the transmission is on.

2.6.3 Phase (or Quadrature) Modulation (PM)

In phase or quadrature modulation, the carrier signal is advanced or retarded in its phase cycle by the modulating base signal. In case of digital modulating signals, the radio carrier signal (at the beginning of each new bit) either will be allowed to retain its phase or will be changed in phase. The advantage of PM/QM is that radio systems using it are relatively less prone to noise interference and other disturbing signals. Thus, lower signal levels at the receiver can be detected; or for similar link lengths, PM receivers have higher availability (lower bit error rates), or for same availability they can operate on longer link lengths.

2.6.4 Quadrature Amplitude Modulation (QAM)

Most modern digital radio systems deploy higher-level modulation schemes based on quadrature amplitude modulation (QAM) schemes where both the amplitude and phase of the carrier signal are modified. The high-level modulation schemes code multiple bits at a time, so that *baud rate* of the transmitted radio signal may be reduced. The *baud rate* or *symbol rate* is the rate at which the radio receiver must be able to distinguish the *symbols* of the incoming signal. The higher the baud rate of the signal, the greater the agility required at the receiver and transmitter.

QAM modulation leads to very sensitive receivers and various higher modulation versions of QAM are available which have the advantage of allowing increased number of bits per symbol to be transmitted. Higher level modulation schemes are built as modem constellations by combining different amplitude levels and phase shift levels.

The 8-QAM modem constellation illustrated in Figure 2.14 combines two levels of amplitude (low and high) and four levels of phase shifts (0, 90, 180, and 270 degrees) leading to eight different states that allows three bit combinations to be represented by one state or symbol consisting of an amplitude and a phase shift combination.

An alternate representation of the 8-QAM modulation scheme is shown in Table 2.2. Even higher modulation schemes like 16-, 32-, 64-, 128-, and 256-QAM

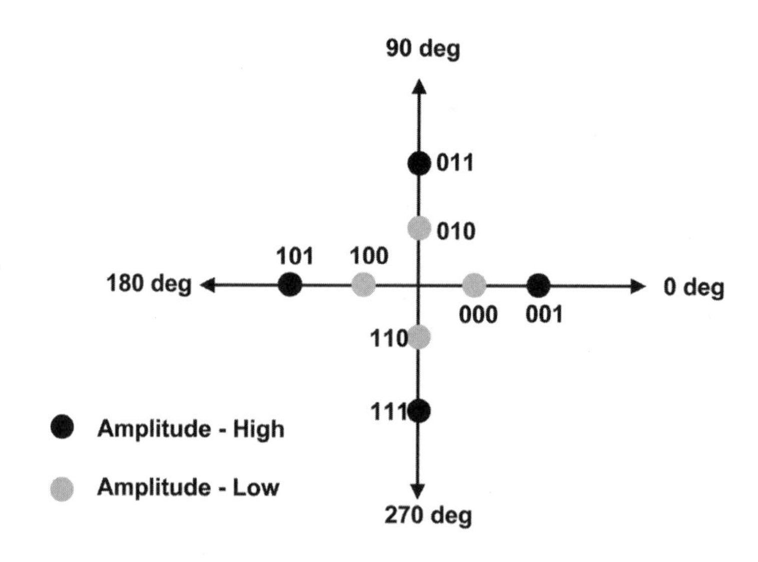

Figure 2.14. Modem constellation diagram for 8-QAM modulation.

TABLE 2.2. Bit Combinations and Signal attributes for 8-QAM Modulation

Bit Combination	Signal Amplitude	Phase Shift
000	Low	000 degrees
001	High	000 degrees
010	Low	090 degrees
011	High	090 degrees
100	Low	180 degrees
101	High	180 degrees
110	Low	270 degrees
111	High	270 degrees

schemes are possible (and are frequently deployed) by combining appropriate amplitude levels and phase shifts. Though use of higher-level modulation schemes like 64-QAM increases the number of bits/symbol (or state), thereby contributing toward more efficient use of the available frequency spectrum, they also introduce higher levels of signal interference and complexity in the radio system design. It has been found that the optimum QAM modulation level that achieves the best possible capacity over a geographic area using a limited amount of spectrum is the 8-QAM scheme. The general conclusion on the use of higher modulation schemes include:

- Use of higher modulation schemes increases the bits/Hz of radio bandwidth.
- Radio systems using higher modulations are more susceptible to interference.
- Higher modulation levels may reduce the possible range (geographic coverage) due to loss from interference.
- Higher modulation schemes are more effective in increasing capacity where the base stations are isolated from each other and there are no nearby sources of interference.

CHAPTER 3

A Refresher on Cellular Mobile and Cordless Telecommunication Systems

Many implementations of WLL systems utilize either the cellular mobile systems or low-power cordless telecommunication systems as their underlying basis. In fact, many vendors of cellular and/or cordless telecommunication systems have developed WLL systems that are essentially stripped-down versions of their cellular or cordless systems. This chapter therefore provides a systems-level overview of the radio and network aspects of the most commonly deployed cellular and cordless telecommunication systems, which include:

- GSM, the European TDMA/FDD cellular standard
- IS-136, the North American TDMA/FDD cellular standard
- IS-95, the North American CDMA/FDD cellular standard
- DECT, the European TDMA/TDD cordless telecommunications standard
- PHS, the Japanese TDMA/TDD cordless telecommunications standard

This chapter also contains a brief introduction to cellular mobile system enhancements and evolution to the third-generation (3G) mobile communication systems, which have started to be deployed in some countries. WLL systems are now being developed using these 3G radio technologies.

3.1 THE CELLULAR CONCEPT

The cellular concept was developed and first introduced by the Bell Laboratories in the early 1970s. One of the most successful initial implementations of the cellular concept was the advanced mobile phone system (AMPS), which has been deployed in the

The contents of Chapter 3 are mainly derived from R. Pandya, *Mobile and Personal Communication Systems and Services*, IEEE Press, Piscataway, NJ (2000).

Introduction to WLLs. By Raj Pandya
ISBN 0-471-45132-0

United States since 1983. The principle of cellular systems, as initially proposed by Bell Laboratories, is to divide a large geographic service area into cells with diameters from 2 to 50 km, each of which is allocated a number of radio-frequency (RF) channels. Transmitters in each adjacent cell operate on different frequencies to avoid interference. However, transmit power and antenna height in each cell is relatively low so that cells that are sufficiently far apart can reuse the same set of frequencies without causing co-channel interference. Figure 3.1 illustrates an idealized view of a cellular mobile system where cells are depicted as perfect hexagons.

As the demand for cellular mobile service grows, additional cells can be added; and as traffic demand grows in a given area, cells can be split to accommodate the additional traffic. The theoretical coverage range and capacity of a cellular system is therefore unlimited.

One of the key requirements of a cellular system is that it should provide the capability to hand off calls in progress, as the mobile terminal/user moves between cells during the call. As far as possible, the handoff should be transparent to the user in terms of interruption and/or call failure. Handoff between channels (in the same cell) may also be required for reasons like load balancing, emergency call handling, and meeting requirements on transmission quality.

Following the development of the cellular concept, a number of cellular mobile systems using FDMA/FDD channels were developd and deployed. Examples of these first-generation (1G) mobile systems include advanced mobile phone system (AMPS), Nordic mobile telecommunication (NMT), and total access communication system (TACS). These have now been replaced almost entirely by the large-scale deployment of second-generation (2G) digital cellular mobile and cordless telecommunication systems. Whereas the digital cellular mobile systems utilize

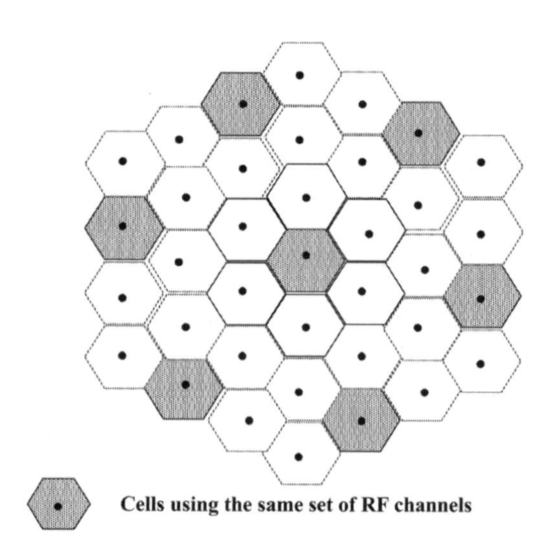

Figure 3.1. Concept of frequency reuse in cellular mobile systems.

TDMA/FDD (GSM, IS-136) or CDMA/FDD (IS-95) channels, cordless telecommunication systems tend to deploy TDMA/TDD (DECT, PHS) techniques.

3.2 BASIC CELLULAR SYSTEM ARCHITECTURE AND OPERATION

3.2.1 Cellular System Architecture

As mentioned earlier, the coverage area of a cellular system is partitioned into a number of smaller areas or *cells*, with each cell served by a base station (BS) for radio coverage. The BSs are connected through fixed links to a base station controller (BSC) that connects to a mobile switching center (MSC). MSC is a local switching exchange with additional features to handle mobility management requirements of a cellular system. In order to handle the dynamic nature of terminal location information and subscription data, the MSC interacts with appropriate databases that maintain subscriber data and location information. The MSC also interconnects with the PSTN, because a significant number of calls in a cellular mobile system either originate from or terminate at fixed network terminals or use the fixed network as a transit network. Figure 3.2 illustrates the basic components of a typical cellular mobile system.

Based on the frequency spectrum made available by the licensing authority and the cellular standard in use, the cellular system is able to define a number of radio channels for use across its serving area. The available radio channels are then partitioned into groups of channels, and these groups of channels are allocated to

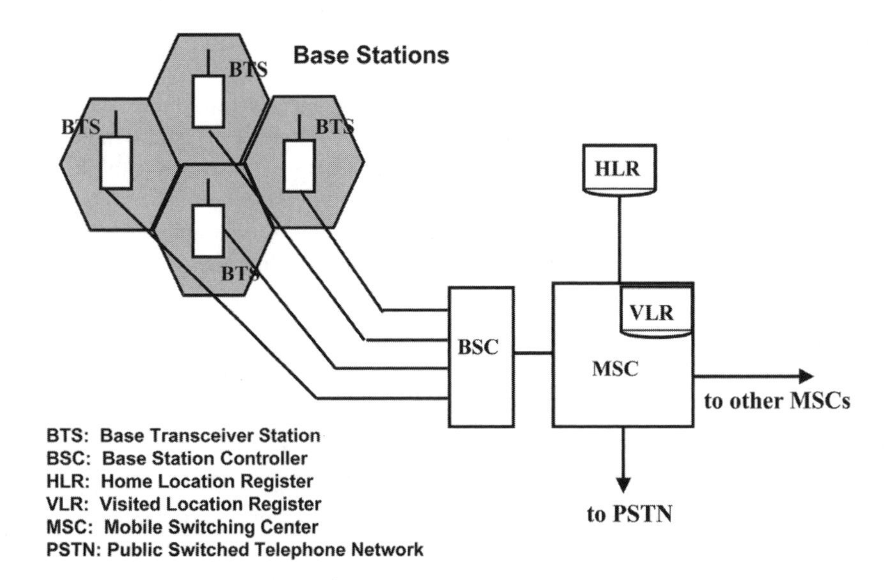

Figure 3.2. Basic components of a typical cellular mobile systems.

individual cells forming the entire serving area. Individual channels or a particular group of channels can be reused in cells that are located far enough apart. A key feature of radio system planning activity consists of designing the cell sizes, their locations, and allocation of radio channels to individual cells. Whereas definition of channels for assignment to individual cells within a cellular system is straightforward in the case of systems based on FDMA and TDMA methods, systems based on CDMA technique require a different view on what constitutes a radio channel in this context.

In each cell, one radio channel is set aside for carrying signaling information between the network (i.e., the BS) and the mobile stations in that cell. The signaling channel is used in the mobile-to-BS direction to carry signals for location updating, mobile originated call setup, and responses to incoming call setup messages (e.g., paging response). In the reverse direction (BS-to-mobile) the signaling channel carries messages related to operating parameters (e.g., location area identity, cell identity), call setup (e.g., paging), and location updating.

3.2.2 Location Updating and Call Setup

In order to deliver an incoming call to a mobile station, it is necessary that the network (i.e., the MSC and the associated location database) maintains information on the location of a mobile station as it moves through the coverage area. The mobile station monitors the overhead information broadcast by the network on the signaling channel and updates the operating parameters as necessary. It also checks the location information (e.g., location area identity) broadcast by the new cell; and if it differs from the previous cell, the mobile informs its new location to the network which updates its location register(s). The information is then used to route incoming calls to the MSC currently serving the mobile and for determining the paging broadcast area for the mobile.

The exact procedures for mobile originated and mobile terminated call setup depend on the technical standard deployed in a particular mobile system. In general, however, the procedures described here apply in most cases. A mobile user originates a call by keying in the called number and depressing the *send* key. The mobile transmits an access request on the up-link signaling channel. If the network can process the call, the BS sends a speech channel allocation message that enables the mobile to lock on the designated speech channel allocated to that cell while the network proceeds to set up the connection to the called party. A terminal validation procedure may also be invoked as part of the originating call setup to ensure that the terminal originating the call is a legitimate terminal.

For a mobile terminated call, the network first establishes the current location area for the called mobile through signaling between the home location register (HLR) and the visiting location register (VLR). This process allows the call to be routed to the current serving MSC. The serving MSC initiates a paging message over the down-link signaling channel toward cells contained in the appropriate paging area. If the mobile is turned on, it receives the page and sends a page response to its nearest BS on the signaling channel. The BS sends a speech channel allocation message to the mobile station and informs the network so that the two halves of the connection can be completed.

3.2.3 Handoff and Power Control

During a call, the serving BS monitors the signal quality/strength in terms of carrier-to-interference ratio (C/I ratio) from the mobile. If the signal quality/strength falls below a predesignated threshold, the network requests the neighboring BSs to measure the signal quality from the mobile. If another BS indicates better signal quality/strength than the serving BS, a signaling message is sent to the mobile on the speech channel (using a *blank and burst* procedure) from the current BS asking the mobile to retune to a free channel in the neighboring cell. The mobile retunes to the new channel (in the new cell), and simultaneously the network switches the call to the new BS. The signal quality measurements and new cell selection generally takes several seconds, but the change of speech channels (handoff) is essentially transparent to the user except for a very brief break in transmission in FDMA- or TDMA-based systems. A typical intercell handoff situation is shown in Figure 3.3.

Generally, the size of cells within a given cellular system may vary from a radius of 1 km in the inner city to more than 20 km in a rural area. It is therefore not necessary for the mobile station to transmit at full power at all times to maintain satisfactory signal level at the BS. Most FDMA and TDMA cellular standards therefore provide for a feature where the BS signals the mobile to operate at one of a series of transmit power levels depending on the distance between the mobile and the BS antenna. The main reason for such a feature is to minimize co-channel interference and to conserve terminal battery power. In CDMA-based cellular systems, accurate power control is essential for avoiding the so-called *near–far* problem and proper operation of the system. Power control issues associated with CDMA systems are addressed in Section 3.6.4.3.

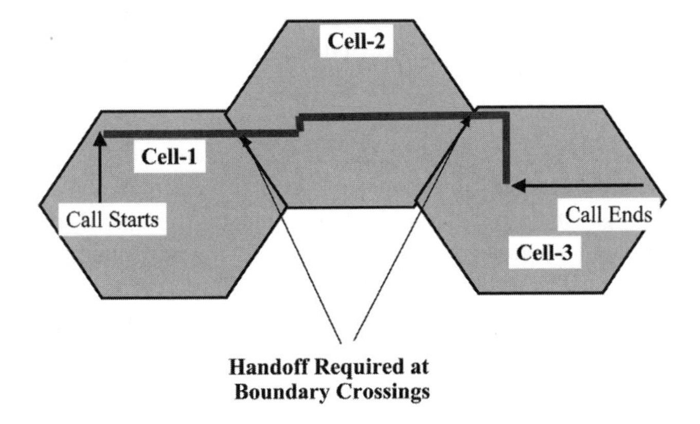

Figure 3.3. A typical intercell handoff in a cellular mobile system.

3.3 INTRODUCTION TO DIGITAL CELLULAR MOBILE SYSTEMS

Concepts of digital radio were first deployed in military applications to provide improved reception in a highly interference prone environment and to provide a high level of security (through encryption) against eavesdropping on the radio path and against unauthorized access (through strong authentication and verification procedures). The driving force behind commercial digital cellular systems is not only the increased system capacity but also the reduction of mobile terminal size and the average power requirements, which, in turn, increases the subscriber terminal battery life and reduces the terminal cost. The development of low-rate codecs, the dramatic increase in the device densities of integrated circuits, and the advances in digital signal processing have made completely digital (second generation) cellular mobile systems a resounding commercial success.

Besides the Japanese PDC digital cellular system whose deployment is limited to Japan, there are two major digital cellular standards based on TDMA/FDD: the European GSM (global system for mobile communications) standard and the North American IS-136 or DAMPS (digital AMPS) standard. Whereas the former is a stand-alone replacement to analog systems, the latter is designed to coexist with the analog (AMPS) system using dual-mode terminals. The GSM system and its frequency up-shifted version called DCS1800 has so far been adopted by a large number of operators worldwide and has captured the largest global subscriber base among the current digital cellular mobile systems. Deployment of cellular mobile systems based on North American IS-136 standard is essentially restricted to North and South America where they coexist with AMPS analog systems.

A CDMA-based cellular mobile system was standardized in the United States as TIA IS-95. CDMA cellular systems claim to provide a significant capacity and cost advantage over TDMA-based systems. CDMA systems based on the IS-95 standard are now in commercial operation in North and South America as well as in many Asian countries like Japan, Korea, and China.

3.4 GSM: THE EUROPEAN TDMA DIGITAL CELLULAR MOBILE STANDARD

The GSM standard was developed by the *Groupe Speciale Mobile*, which was an initiative of the Conference of European Post and Telecommunications (CEPT) administrations. The underlying aim was to design a uniform pan-European mobile system to replace numerous incompatible analog systems in use in different European countries. The responsibility for GSM standardization work now resides with ETSI (European Telecommunication Standards Institute). The main characteristics and capabilities associated with the initial GSM standard included:

- Fully digital system utilizing the 900-MHz frequency band
- TDMA/FDD access over radio carriers with 200-kHz carrier spacing

- Eight full-rate or 16 half-rate TDMA channels per carrier
- User/terminal authentication for fraud control
- Encryption of speech and data transmissions over the radio path
- Full international roaming capability
- Low-speed data services (up to 9.6 kb/s)
- Compatibility with ISDN for supplementary services
- Support for short message service (SMS)

The GSM standard has been undergoing continuous extensions and enhancements to support more services and capabilities like high-speed circuit switched data (HSCSD), general packet radio service (GPRS), and customized applications for mobile network enhanced logic (CAMEL). A version of GSM operating in the 1900-MHz band (PCS1900) has also been standardized in North America and is deployed in many markets.

3.4.1 GSM Reference Architecture and Function Partitioning

As shown in Figure 3.4, the GSM system is comprised of base transceiver stations (BTSs), base station controllers (BSCs), mobile switching centers (MSCs), and a set of registers (databases) to assist in mobility management and security functions. All signaling between the MSC and the various registers (databases) as well as between MSCs takes place using the signaling system 7 (SS7) network with the application level messages using the mobile application protocol (MAP) designed specifically

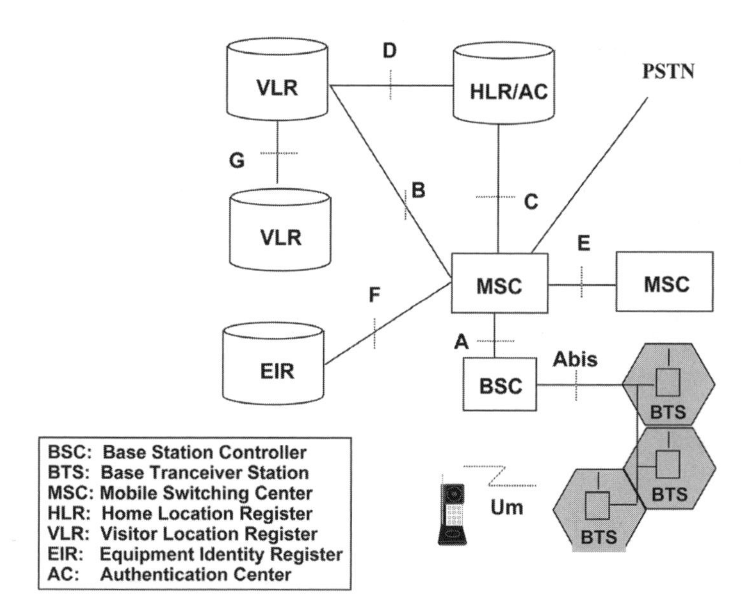

Figure 3.4. Reference architecture and signaling interfaces for GSM.

for GSM. The MAP protocol utilizes the lower layer functions from the SS7 protocol stack. These include MTP (message transfer part), SCCP (signaling connection control part), and TCAP (transaction capability application part). The signaling between the MSC and the national PSTN/ISDN is based on the national options for SS7 telephone user part (TUP) or ISDN user part (ISUP). The base signaling transport for the A interface is also SS7.

3.4.1.1 Base Station System (BSS) This comprises a base station controller (BSC) and one or more subtending base transceiver stations (BTSs). The BSS is responsible for all functions related to the radio resource (channel) management. This includes the management of radio channel configuration with respect to their use as speech, data, or signaling channels, allocation and release of channels for call setup and release, control of frequency hopping, and transmit power at the mobile station (MS). Range of functions performed by the BSS include:

- Radio resource control
- Frequency hopping and power control
- Handoff management
- Digital signal processing

3.4.1.2 Mobile Switching Center (MSC) for GSM This can be viewed as a local ISDN switch with additional capabilities to support mobility management functions like terminal registration, location updating, and handoff. Furthermore, unlike a local switch in a fixed network, the MSC does not contain the mobile subscriber parameters (which are dynamic and are held in the VLR associated with the MSC). Thus, the major functions performed by the MSC include:

- Call setup, supervision, and release
- Digit collection and translation
- Call routing
- Billing information collection
- Mobility management (terminal registration, location updating, and handoffs)
- Paging and alerting
- Management of radio resources during a call
- Echo cancellation
- Connection management to BSS, other MSCs, and PSTN/ISDN
- Interrogation of appropriate databases (HLR, VLR)

3.4.1.3 Home Location Register (HLR) This represents a centralized database that has the permanent data fill for the mobile subscribers in a large service area (generally one per GSM network operator). It is referenced using the SS7 signaling capabilities for every incoming call to the GSM network for determining the current location of the subscriber—that is, for obtaining the mobile station routing number (MSRN) so that the call may be routed to the mobile station's serving MSC. The HLR is kept updated with the current locations of all subtending mobile subscribers including those who may have roamed to another network operator within or outside the country. The routing information is obtained from the serving VLR on a call-by-call basis so that for each incoming call the HLR queries the serving VLR for a MSRN.

Usually, one HLR is deployed for each GSM network for administration of subscriber configuration and service. Besides the up-to-date location information for each subscriber which is dynamic, the HLR maintains the following subscriber data on a permanent basis:

- International mobile subscriber identity (IMSI)
- Service subscription information
- Service restrictions
- Supplementary services (subscribed to)
- Mobile terminal characteristics
- Billing/accounting information

3.4.1.4 Visiting Location Register (VLR) This represents a temporary data store, and generally there is one VLR per MSC. It contains information about the mobile subscribers who are currently in the service area covered by the MSC/VLR. The VLR also contains information about locally activated features such as call forward on busy. Thus, the temporary subscriber information resident in a VLR includes:

- Features currently activated and available to the subscriber
- Temporary mobile station identity (TMSI)
- Current location information about the MS (e.g., location area and cell identities)

3.4.1.5 Authentication Center (AC) This is generally associated with the HLR, and it contains authentication parameters that are used on initial location registration, on subsequent location updates, and on each call setup request from the MS. In the case of GSM, the AC maintains the authentication keys and algorithms, and it provides the security triplets (RAND, SRES, and Kc) to the VLR so that the user authentication and radio channel encryption procedures may be carried out within the visited network. The authentication center for GSM contains the security modules for the authentication keys (Ki) and the authentication and cipher key generation algorithms A3 and A8, respectively.

3.4.1.6 Equipment Identity Register (EIR) This maintains information to authenticate terminal equipment so that fraudulent, stolen, or non-type-approved terminals can be identified and denied service. The information is in the form of white, gray, and black lists that may be consulted by the network when it wishes to confirm the authenticity of the terminal requesting service.

3.4.2 GSM Radio Aspects

In GSM the up-link (mobile-to-base) frequency band is 890–915 MHz and the corresponding down-link (base to mobile) band is 935–960 MHz, resulting in a 25-MHz spectrum for duplex operation. The GSM uses time division multiple access (TDMA) and frequency division multiple access (FDMA) whereby the available 25-MHz spectrum is partitioned into 124 carriers (carrier spacing is 200 kHz) and each carrier in turn is divided into eight time slots (radio channels). Each user transmits periodically in every eighth time slot in an up-link radio carrier and receives in a corresponding time slot on the down-link carrier. Thus several conversations can take place simultaneously at the same pair of transmit/receive radio frequencies. The frequency division duplex structure for GSM was illustrated in Figure 2.4. The frame structure used in GSM is illustrated in Figure 3.5, and the radio parameters for GSM are summarized in Table 3.1.

In the GSM system the digitized speech at 64 kb/s is passed through a speech coder (vocoder) which compresses the 64-kb/s PCM speech to a 13-kb/s data rate. The vocoder models the vocal tract of the user and generates a set of filter parameters that are used to represent a segment of speech (20 ms long), and only the filter parameters and the impulse input to the filter are transmitted on the radio interface. The speech coding technique increases the spectral efficiency of the radio interface, thereby increasing the traffic capacity of the system (more users over a limited bandwidth). The linear predictive, low-bit rate (LBR) vocoder is based on residual pulse excitation–long term prediction (RPE–LTP) techniques. The GSM

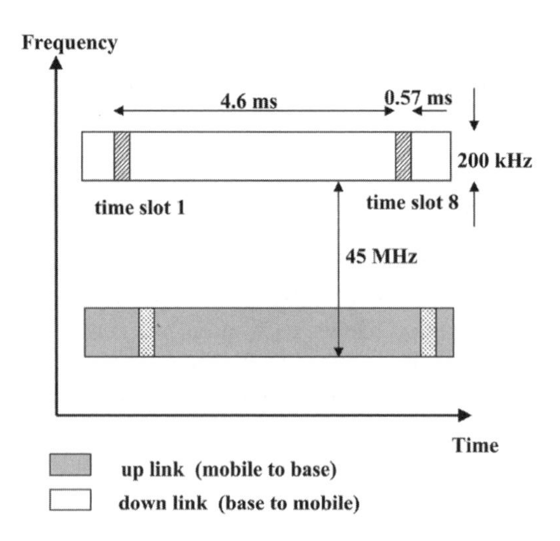

Figure 3.5. Frame and time slot structure for GSM.

TABLE 3.1. Typical Radio Parameters for GSM

System Parameter	Value (GSM)
Multiple access/duplex method	TDMA/FDMA/FDD
Up-link frequency (mobile-to-base)	890–915 MHz
Down-link frequency (base-to-mobile)	935–960 MHz
Channel bandwidth	200 kHz
Number of duplex frequency carriers	124
Channels/carrier	8 (full rate), 16 (half rate)
Frame duration	4.6 ms
Interleaving duration	40 ms
Modulation	GMSK[a]
Speech coding method	RPE–LTE[b] convolutional
Speech coder bit rate	13 kb/s (full rate)
Associated control channel	Extra frame
Handoff scheme	Mobile-assisted
Mobile station power levels (W)	0.8, 2, 5, 8

[a]Gaussian minimum shift keying.
[b]Residual pulse excitation–long-term prediction.

vocoder also permits the detection of silent periods in the speech sample, during which transmit power at the mobile station can be turned off to save power, extend battery life, and also reduce interference.

The vocoder output is error protected by passing it through a channel encoder that utilizes both a parity code and a convolution code. The source-encoded data are then interleaved to combat the effects of burst errors on the radio path. The interleaved data are then modulated using GMSK (Gaussian minimum shift keying) and passed through a duplexer that provides filtering to isolate transmit and receive signals. This process of information transfer for speech signals, in the up-link direction across the radio interface, is illustrated in Figure 3.6. A similar process is applicable for the down-link.

Whereas the above digital signal processing in GSM radio leads to higher capacity and better speech signal quality, it also introduces additional delays. Therefore echo control devices become necessary for all GSM calls that use the PSTN as a transit network. An echo canceller is included at the PSTN interface of the GSM network in order to remove any echoes being returned from the PSTN to the GSM mobile user.

3.4.2.1 Frequency Hopping Frequency hopping capability on the radio interface is provided in GSM in order to reduce the effects of multipath fading when a mobile station is stationary or moving slowly. Frequency hopping also provides added security against unauthorized eavesdropping on a call in progress. Frequency hopping is achieved in such a way that, according to the calculated sequence (based on a simple algorithm) on both sides of the air interface, the MS sends and receives

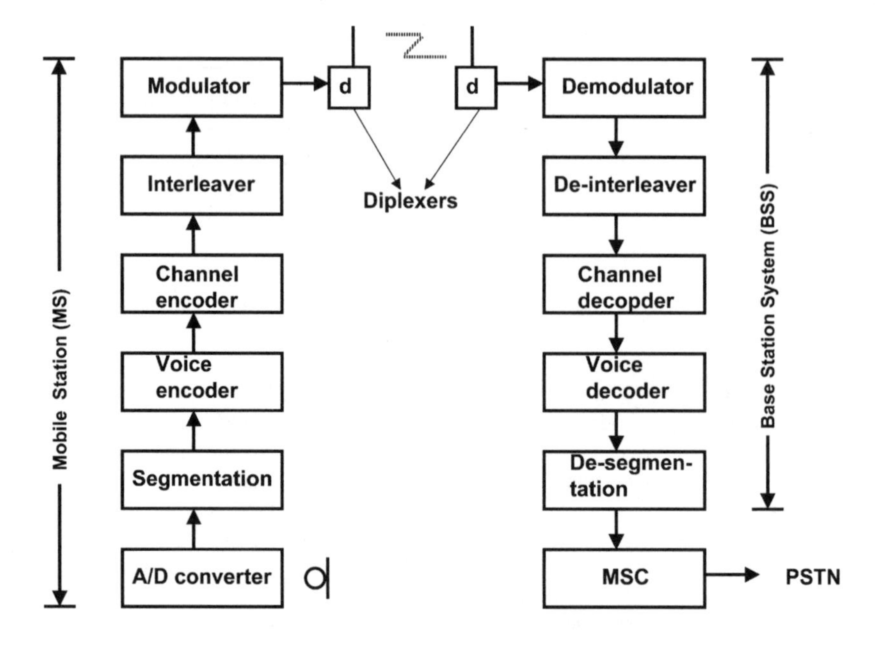

Figure 3.6. Typical radio system components for GSM.

each time slot (burst) on a different frequency. Frequency hopping is applied only on the traffic and nonbroadcast channels in order to mitigate interference from nearby sources.

3.4.2.2 Discontinuous Transmission (DTx) or Voice-Activated Transmission This is utilized in GSM in order to maximize spectrum efficiency. The technique is based on detecting voice activity and switching on the transmitter during only those periods when there is active speech to transmit. Furthermore, switching off the transmitter during silent periods also reduces interference in the air, thereby allowing the use of smaller-frequency reuse clusters and reduces battery power consumption in the mobile terminal. Up to 50% gain in spectrum efficiency can be achieved through DTx. An adaptive threshold voice activity detector algorithm is used in the GSM radio equipment to detect silent periods and interrupt transmission. At the receiving end the DTx is detected and empty frames are filled with comfort noise. The voice activity detector algorithms and its implementation need to be carefully designed in order to minimize speech clipping and the resulting degradation in speech quality.

3.4.3 Security Aspects for GSM

The security procedures in GSM are aimed at protecting the network against unauthorized (fraudulent) access and protecting the privacy of the mobile subscribers against eavesdropping on their communications. The security procedures also prevent the possibility of tracing the identity and location of the subscribers as they roam within or outside the home network. In GSM, protection from unauthorized access is achieved through strong authentication procedures that validate the true identity of the subscriber before he/she is permitted to receive service. Eavesdropping on the subscribers' communications is prevented by ciphering the information channel across the radio interface (i.e., applying encryption on the digital stream on the radio path). In order to protect the identity and location of the subscriber, the appropriate radio signaling (control) channels are also ciphered, and a temporary mobile subscriber identity (TMSI) is used over the radio path instead of the actual international mobile subscriber identity (IMSI). Note that the privacy mechanisms (encryption and use of TMSI) are used only over the radio path and not within the fixed infrastructure where the communications are transmitted in the clear, as they are in PSTN/ISDN.

In GSM systems, each mobile user is provided with a subscriber identity module (SIM). Two possible versions of SIM are defined in GSM standard: One version is a chip card the size of a credit card to be inserted in the mobile terminal; the other version is a small (25 mm × 15 mm) plug-in SIM that can be installed in the mobile terminal on a semipermanent basis. At the time of service provisioning, the IMSI and the appropriate security parameters (for authentication and encryption) are programmed into the SIM by the GSM operator. The authentication procedure used in GSM is a private key authentication method that utilizes a challenge–response mechanism.

An additional security feature in GSM is the equipment identity register (EIR), which maintains black, gray, and white lists of international mobile equipment identities (IMEIs) for monitoring mobile equipment. Each mobile terminal is assigned a unique IMEI that consists of a type approval code, a final assembly code, and a serial number. The IMEI is used to validate mobile equipment (terminals) so that non-type-approved, faulty (not meeting the spurious emission regulations) or stolen terminals are denied service.

3.4.4 GSM Protocol Model

As shown in Figure 3.7, the signaling at the radio interface (Um) consists of LAPDm at Layer 2. LAPDm is a modified version of LAPD (link access protocol for D channel), used in ISDN user–network access, to accommodate radio interface-specific features. Layer 3 is divided into three sublayers that deal with radio resource (RR) management, mobility management (MM), and connection management (CM).The radio resource management is concerned with managing logical channels, including the assignment of paging channels, signal quality measurement reporting, and handoff. The mobility management sublayer provides functions necessary to support user–terminal mobility which include terminal registration, terminal location updating, authentication, and IMSI detach/attach. The connection management sublayer is concerned with call and connection control, establishing

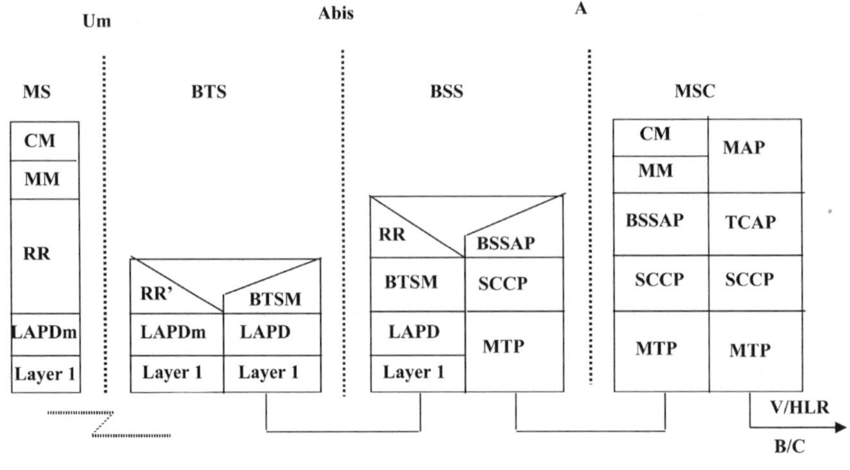

LAPD: Link Access Protocol for D channel
RR: Radio Resource (management)
MM: Mobility Management
CM: Connection Management
BTSM: BTS Management

BSSAP: BSS Application Protocol
MTP: Message Transfer Part
SCCP: Signaling Connection Control Part
TCAP: Transaction Capability Application Part
MAP: Mobile Application Protocol

Figure 3.7. Protocol reference model for GSM.

and clearing calls/connections, management of supplementary services, and support of the short message service.

The Abis interface between the BTS and the BSC, though fully defined in the GSM standard, tends to be a proprietary interface in most GSM implementations. The radio resource layer (RR′) in the BTS is responsible for channel establishment and release, handoff, and paging. The BTS management (BTSM) layer looks after the management of all aspects of the radio channels including radio link layer, control channels, and transceiver management.

The BSC to MSC interface (A interface) and the interfaces between a MSC and a V/HLR or another MSC deploy ITU-T signaling system 7 and the mobile application protocol (MAP). Connections between the MSC and other PSTN/ISDN exchanges utilize the SS7 telephone user part or the ISDN user part specific to the country or region where the GSM system is deployed.

3.5 IS-136: THE NORTH AMERICAN TDMA DIGITAL CELLULAR MOBILE STANDARD

The key objectives for the GSM standard in Europe were to specify a single, high-capacity digital cellular system for Europe to replace the existing, incompatible analog systems that inhibited roaming across Europe. The GSM was also assigned a new frequency spectrum in the 900-MHz frequency band for this purpose.

The primary aim of the digital cellular standards in North America was to increase the capacity of the existing spectrum and to provide a wider range of services with improved performance. As such, the digital system was required to coexist with the analog AMPS system. Two digital cellular systems were standardized in North America: One is based on TDMA/FDD, namely, the TIA/EIA IS-136 standard (also referred to as DAMPS); the other based on CDMA/FDD, namely, the TIA/EIA IS-95 standard. These digital cellular standards require that both the AMPS analog system and the new digital systems coexist (using dual mode terminals sharing the common 800-MHz frequency band), with the expectation that digital system(s) will eventually replace the analog AMPS system.

3.5.1 Network Reference Model

Figure 3.8 presents the functional entities and the associated interface reference points for the North American digital cellular system. The network reference model is the same for both the TDMA-based (IS-136) and CDMA-based (IS-95) radio systems. The model is used as a basis for specifying the messages and protocols for intersystem operation based on the ANSI-41 standard. The distribution of various functions within the physical network entities can vary among different implementations (e.g., the VLR functions are generally collocated with the MSC).

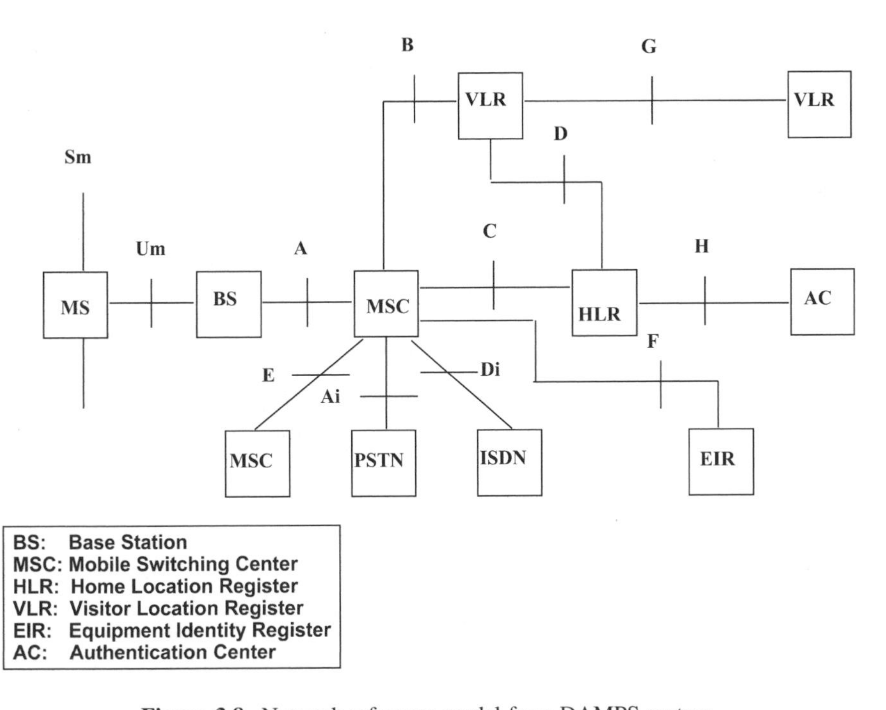

Figure 3.8. Network reference model for a DAMPS system.

The reference architecture and interface reference points in Figure 3.8 are almost identical to those for GSM (Figure 3.4). The functions assigned to each of the entities like BS, MSC, HLR, VLR, AC, and EIR essentially support the same functions as in GSM which were described in Section 3.4. Note that in GSM, the base station is further partitioned into the base station controller (BSC) and base transceiver stations (BTS) with a defined (but rarely implemented) Abis interface.

Notwithstanding these similarities in the reference architecture and interface reference points, the underlying protocols and messages deployed in the DAMPS for intersystem operation are quite different from the GSM. For DAMPS, interfaces B, C, D, and E are fully defined in IS-41 (Revision C), and an SS7-based A interface (similar to the GSM A interface) has also been standardized.

3.5.2 Radio Aspects for DAMPS

DAMPS will utilize the currently allocated spectrum for Analog AMPS—that is, a total of 50 MHz in the 824–849 MHz (up-link) and 869–894 MHz (down-link) with each frequency channel being assigned 30-kHz spacing. Each frequency channel then is time-multiplexed with a frame duration of 40 ms which is partitioned into six time slots of 6.67-ms duration. The frame and time-slot structure used in DAMPS is shown in Figure 3.9, and the associated radio parameters are summarized in Table 3.2.

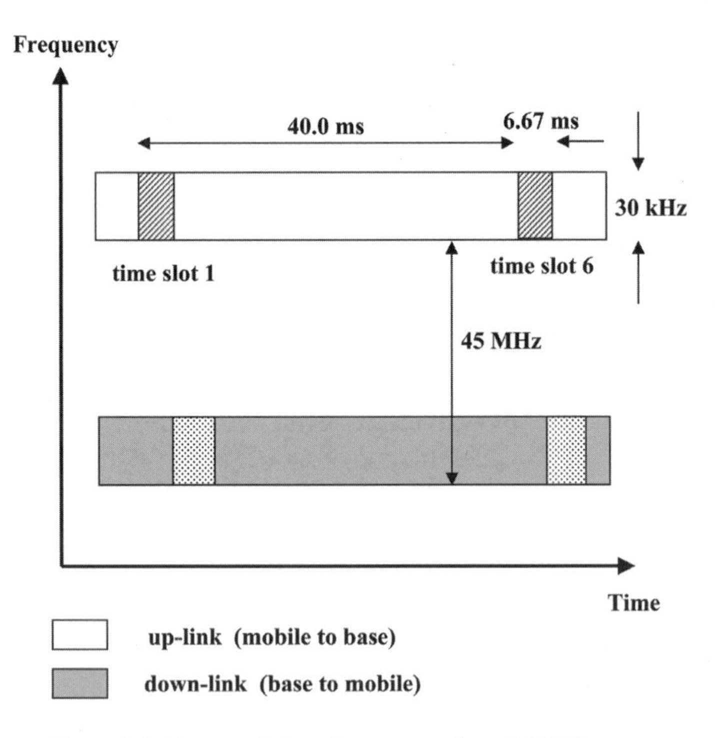

Figure 3.9. Frame and time-slot structure for a DAMPS system.

TABLE 3.2. Typical Radio Parameters/Characteristics for DAMPS

System Parameter	Value (IS-136/DAMPS)
Multiple access/duplexing method	TDMA/FDD
Up-link frequency (mobile-to-base)	824–849 MHz
Down-link frequency (base-to-mobile)	869–894 MHz
Channel bandwidth	30 kHz
Number of channels	832
Time slots/frame	6
Frame duration	40 ms
Interleaving duration	27 ms
Modulation	pi/4DQPSK[a]
Speech coding method	VSELP[b] convolutional
Speech coder bit rate	13.2 kb/s (full rate)
Associated control channel	Same frame
Handoff scheme	Mobile-assisted
Mobile station power levels (W)	0.8, 1.0, 2.0, 3.0

[a]Differential encoded quadrature phase shift keying.
[b]Vector sum excited linear predictive.

For mobile-to-base communication in DAMPS system using a full-rate codec, three mobiles transmit to a single base station radio by sharing the same frequency. This is accomplished by each mobile transmitting periodic bursts of information to the base station in a predetermined order—that is, mobile 1 on time slots 1 and 4, mobile 2 on time slots 2 and 5, and mobile 3 on time slots 3 and 6. When the half-rate codec is implemented, six mobiles will be able to share a frequency channel by each transmitting sequentially on the six available time slots, thereby doubling the capacity.

In the base-to-mobile communication link, each of the three mobiles (full-rate codec case) again shares the same (down-link) frequency channel. Each mobile receives the same information from the base station in the form of a continuous stream of time slots rather than periodic slots (bursts) of information. This continuous transmission is necessary in order for the dual-mode terminals to correctly synchronize to the TDMA transmission without the need for high-accuracy synthesizers (clock) in the mobiles. Each mobile extracts or reads only its assigned time slot(s).

The DAMPS radio system is illustrated in Figure 3.10. The digitized speech at 64 kb/s is passed through a VSELP (vector sum excited linear predictive) coder. The source coder reduces the data rate to 7.95 kb/s, which passes through a channel coder and interleaver. This increases the bit rate to 13 kb/s, which includes error correction and detection bits, control channel data, training sequence data, and guard bits. The channel-coded and interleaved bit stream is then input to the pi/4DQPSK (pi/4-shifted differential quadrature phase shift keying) modulator which modulates the data stream to a carrier in the 800-MHz band.

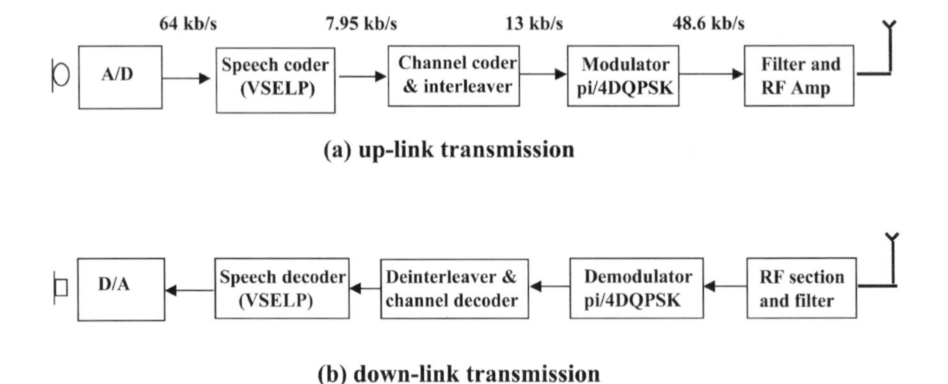

Figure 3.10. Basic components of radio system for DAMPS.

3.5.3 Security Aspects in DAMPS

As in the case of GSM, DAMPS also uses a challenge response procedure based on private key for authentication and privacy. However, the authentication and privacy algorithm, the nature of the private key, and the procedures for generating and transporting the authentication results for verification are different from those used in the GSM. The DAMPS system also supports the option of encryption over the radio channel.

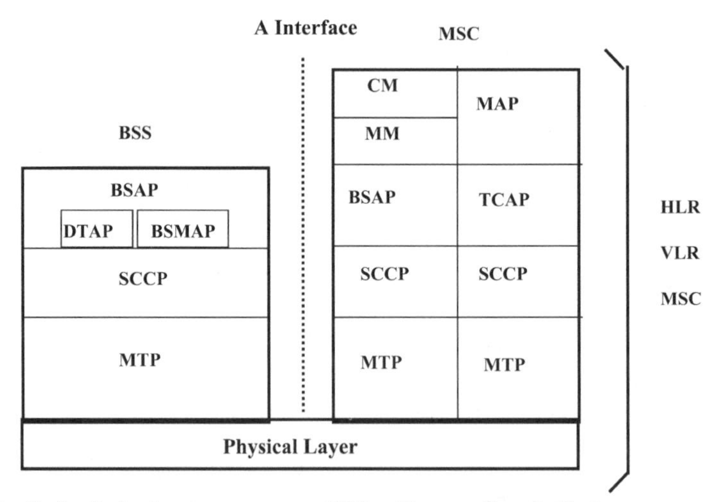

Figure 3.11. Simplified protocol reference model for DAMPS.

3.5.4 Protocol Reference Model for DAMPS

Whereas the GSM specifications provide complete definitions of the Abis interface (between BTS and BSC) and the A interface (between the BSC and the MSC), until recently these interfaces were not defined for DAMPS. The A interface based on the SS7 signaling transport has now been specified, but currently there are no plans to specify an interface equivalent to Abis in GSM. The simplified protocol reference architecture for DAMPS is shown in Figure 3.11.

Functionally, the protocol model for the A interface and the network interfaces (e.g., between the MSC and the HLR) are very similar in that they utilize the same SS7 layers for signaling transport. However, such SS7 entities as the MTP, SCCP, TCAP, and so on, used in DAMPS specifications are based on the SS7 standards in the United States (i.e., ANSI standards for MTP, SCCP, and TCAP) as opposed to the ITU-T standards (used in the GSM system). Furthermore, the mobile application protocol (MAP) used in DAMPS was developed by the TIA as part of ANSI-41 standard to facilitate cellular system interoperability and roaming.

3.6 IS-95: THE NORTH AMERICAN CDMA DIGITAL CELLULAR MOBILE STANDARD

One of the targets set by the CTIA (Cellular Telecommunication Industries Association) for the digital cellular systems for North America was to achieve a 10-fold increase in traffic capacity over the analog AMPS system. The capacity increase provided by the TDMA-based DAMPS does not meet this objective. In 1990, Qualcomm developed and demonstrated a CDMA-based digital cellular system that claimed a 20-fold increase in capacity over the analog AMPS system. The IS-95 standard for the CDMA common air interface was adopted by TIA in 1993, followed by further enhancements to the standard including IS95-A, IS-95B, and ANSI-008 (1900 MHz). CDMA systems based on any of these standards are sometimes referred as *cdmaOne*, which is a trademark of CDMA Development Group.

The services supported by CDMA cellular systems are analogous to those supported by GSM and DAMPS, and its network reference architecture is the same as that for the DAMPS TDMA cellular system (Figure 3.8).

3.6.1 Radio Aspects for an IS-95 CDMA Digital Cellular System

The spread spectrum techniques used in IS-95 CDMA cellular systems are adaptation of similar techniques used extensively in military applications since 1950. In these techniques, the normal payload data are modulated and transmitted using a special spreading code. At the receiving end, the desired signal is recovered by de-spreading the signal with an exact copy of the spreading code in the receiver correlator. Other signals (within the same frequency band) remain fully spread and are perceived as noise.

As illustrated in Figure 3.12, in an IS-95 CDMA system the multiplexing of different information streams (information from multiple users) is achieved by

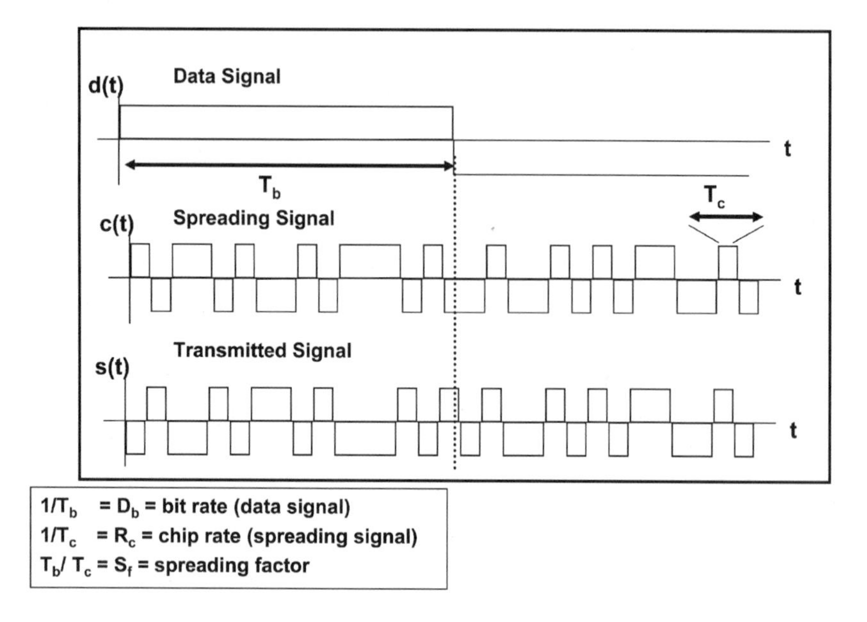

Figure 3.12. Principle of signal spreading in an IS-95 CDMA system.

assigning a code word or code sequence to each information stream. The information stream generated by a user could be the output of a speech coder with a data rate of Tb. Each bit of the user data is multiplied/correlated by the length of the code word or the so-called *spreading code* with a chip rate of 1/Tc, and the resulting output is modulated by a suitable carrier and transmitted. In order to distinguish the bits from the original information stream and the bits from the spreading code, the latter bits are called *chips*.

The choice of the spreading code is very important because the ultimate capability to extract the signal at the receiver is determined by the characteristics of the spreading code sequences used. In order to minimize interference between received signals, it is necessary that the code should have good autocorrelation properties such that when correlated with itself, it generates an impulse like peak at zero offset. Furthermore, the spreading codes should possess the property of orthogonality (i.e., when any two codes in the family are multiplied, the result is zero).

Walsh codes have these properties, but the available number of Walsh codes is generally not sufficient to meet the needs of a mobile communication system with a large number of cells. The next best option that is used in a mobile system is provided by the so-called PN (pseudorandom noise) codes. The PN codes are nearly orthogonal so that the interference between different users is not very severe, and they are not in short supply like Walsh codes.

Each user (terminal) is assigned a distinct PN spreading code to distinguish it from other users (in the same cell/sector). In order to avoid interference from adjacent cells, users in adjoining cells are assigned different PN codes. The unique

PN code sequence assigned to each user terminal is generated by using the terminal's ESN (electronic serial number) as input to a linear shift register (LSR). The base station can then also generate the same PN code sequence (for decoding the incoming signal) by using the ESN as input to a LSR. Forward (base-to-mobile) traffic channels are distinguished using Walsh codes with a length of 64 chips (limit of 64 forward channels per frequency assignment (of 1.23 MHz).

Each base station in the IS-95 cellular mobile system uses the same PN code sequence which is repeated every 27 ms, to encode its signal before transmission. However, each BS applies one of 512 possible delays (time offsets) to the sequence so that the *rake* receiver in the subscriber terminal can distinguish the transmission source (i.e., BS). The rake receiver in the subscriber terminal enables it to distinguish between base stations and to receive its signal from multiple base stations at the same time. This provides space diversity which allows so-called *soft handoffs* in CDMA systems.

3.6.2 Forward Link (Base-to-Mobile) Structure in an IS-95 CDMA System

The IS-95 common air interface (CAI) specifies a forward physical channel (known as forward waveform) design that uses a combination of frequency division, pseudorandom code division and orthogonal signal multiple access techniques. As mentioned earlier, frequency division is achieved by dividing the available cellular spectrum into nominal 1.23-MHz channels by combining 41 30-kHz AMPS channels. These 1.23-MHz channels can be increased from the initial single CDMA channel to multiple channels as demand for digital service increases. An example of a logical forward waveform for an IS-95 CDMA system is shown in Figure 3.13. It consists of a maximum of 64 channels that include a pilot channel, a synchronization channel (optional), paging channels (maximum of 7), and traffic channels.

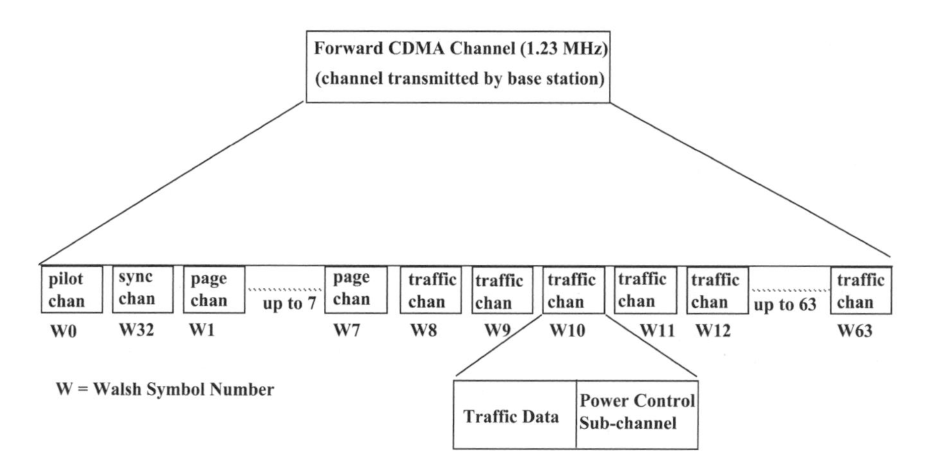

Figure 3.13. Structure of an IS-95 CDMA forward channel.

3.6.2.1 *Pilot Channel* This consists of an unmodulated direct sequence spread spectrum signal with its own identifying spreading code, and it is shared among all users (mobiles) in a sector or cell. Its data content is a sequence of zeros that are not channel-encoded. Each pilot transmits the same spreading sequence at a different time offset which can be used to distinguish signals of different pilots. The basic uses of the pilot channel include:

- Uniquely identifying sectors/cells
- Providing phase/time/signal strength reference
- Identifying multi-path components
- Identifying handoff candidates
- Rescanning or periodically checking for a better sector/cell

Mobiles continuously monitor pilot channels while on a paging or traffic channel so that they can move to a different cell (with better signal).

3.6.2.2 *Synchronisation Channel* This provides a mobile with the system parameters required to synchronize with the network and obtain a paging channel. The synchronization channel therefore includes such information as

- System identification number (SID)
- Network identification number (NID)
- Version of the radio interface being supported
- Precise time-of-day information
- PN offset of the associated pilot channel

3.6.2.3 *Paging Channels* There can be up to seven paging channels on a forward waveform transmitted by a base station, and they contain messages to a specific mobile as well as necessary system parameters. Mobiles are assigned paging channels by a hash function based on the MIN (mobile identification number). Typical messages on the paging channel include pages, traffic channel assignments, and short messages.

3.6.2.4 *Forward Traffic Channels* These are used to transmit voice or data to a mobile that is in a call. There may be up to 63 forward traffic channels depending on the number of paging channels and the presence of synchronization channels. Data rates are flexible (1200, 4800, or 9600 b/s) to support variable rate vocoders, and these are structured in 20-ms frames. Signaling information from the base station to the mobile during a call can be transmitted using *blank and burst* or *dim and burst* methods.

3.6.3 Reverse Channel (Mobile-to-Base) Structure in IS-95 CDMA System

The reverse CDMA channel is used by all the mobiles in a cell coverage area to transmit to the base station. Figure 3.14 shows how the reverse channel appears at the base station receiver, and it is a composite of all the outputs from all the mobiles in the base station's coverage area. At any instant in time, there may be m mobiles engaged in a call (carrying user traffic) and n mobiles trying to gain access to the system. Thus the reverse channel structure allows up to 62 different traffic channels and 32 different access channels. However, considerably fewer than these numbers of channels may be in use at any time.

3.6.3.1 Access Channels An access channel allows a mobile to communicate with the system when it needs to initiate an action like registration or call origination or when it needs to respond to messages received on a paging channel. Since multiple mobiles may attempt access at the same time, the access channel utilizes some form of slotted random access protocol for contention management. The data rate for the access channel is 4.8 kb/s, and the number of access channels in a cell are configurable—generally one or two access channels for each paging channel.

3.6.3.2 Reverse Traffic Channels These are used for transmitting voice or data traffic from a mobile to the base station. Each of these channels is paired with a corresponding forward traffic channel. Signaling information during a call is again transmitted using *blank and burst* or *dim and burst* mode.

Various types of channels used in a CDMA cellular system, and their characteristics are summarized in Table 3.3.

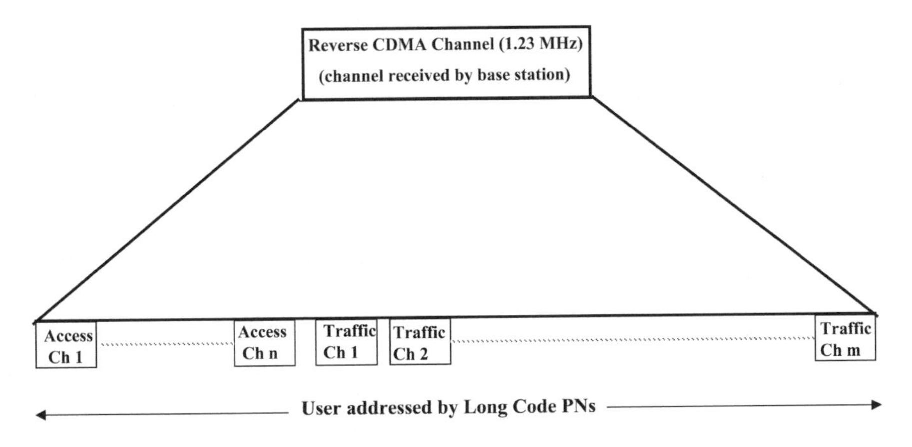

Figure 3.14. Structure of an IS-95 CDMA reverse channel.

TABLE 3.3. Channel Types in IS-95 CDMA Digital Cellular Systems

Channel Type	Application	Quantity	Maximum Rate (kb/s)	Spreading Code
Pilot	System monitoring	1	N.A.	Walsh code 0
Sync.	Synchronization	0 or 1	1.2	Walsh code 32
Paging	Signaling (BS- to-MS)	Up to 7	9.6	Walsh code 1–7
Traffic (forward)	Voice/data (BS-to-MS)	Up to 63	9.6/14.4	Walsh codes 8–31 and 32–63
Access[a]	Signaling (idle MS-to-BS)	Up to 14	4.8	Long code mask
Traffic (reverse)	Voice/data (MS-to-BS)	Up to 63	9.6/14.4	Mobile-specific long code mask

[a]One to two access channels for each paging channel are generally provided.

3.6.4 Some Key Features of IS-95 CDMA Systems

3.6.4.1 Diversity Cellular systems are prone to multipath fading, and some form of diversity method is required to mitigate the effects of fading. The types of diversity that are available in a CDMA system include:

- Time diversity provided by symbol interleaving, error detection, and correction coding
- Frequency diversity provided by the 1.25-MHz wideband signal
- Space (path) diversity provided by dual cell-site receive antennas, multipath rake receivers, and multiple cell sites (soft handoff)

Whereas time and frequency diversity can be provided to certain extent in FDMA and TDMA systems, a unique feature of CDMA systems is their ability to provide extensive path diversity and potential for improved performance in a difficult propagation environment. The path diversity is achieved through multipath processing using parallel correlators for the PN waveforms at the receivers in the mobile and the cell sites. Receivers using parallel correlators (rake receivers) allow signals arriving on individual paths to be tracked independently, and the sum of their received signal strengths is then used to demodulate the signal. The deployment of multiple correlators for the simultaneous tracking of signals from different cells provides the underlying basis for *soft handoff* in CDMA systems.

3.6.4.2 Soft Handoff Soft handoff in a CDMA system results from the systems capability to simultaneously deliver signals to a mobile through more than one cell. Thus the handoff sequence in a CDMA system involves transition from the donor cell to both the donor and the receiving cell and finally to the receiving cell (similar to the handoff of a baton in a relay race). This *make-before-break* procedure not only reduces the probability of a dropped call during handoff, but also makes the handoff virtually undetectable by the user. In this regard, the FDMA (analog)- and TDMA (digital)-based systems utilize a *break-before-make* call switching procedure with potential adverse impact on quality of service. The soft handoff process in the CDMA system is illustrated in Figure 3.15.

The handoff is *mobile-assisted* in that the mobile station with a call in progress continues to monitor the signal strength from neighboring cells; and if the signal from one of these cells becomes comparable to that of the original cell, it initiates the handoff process by sending a control message to the MSC. The MSC responds by establishing a link to the mobile station through the new cell (identified in the control signal from the mobile) while maintaining the old link. While the mobile station is located in the transition region between two adjacent cells, the call can be supported by signals through both cells, thereby eliminating the common *border cell problems* of frequent back-and-forth handoffs between the two cells. The original cell will hand off the call only when the mobile station is firmly established in the new cell.

3.6.4.3 Power Control In a CDMA system the mobile-to-base station link is subject to the so-called *near–far problem* whereby a mobile station close to the base station has a much lower path loss than do mobiles that are far removed from the base. If all the mobile stations were to use the same transmit power, the mobile(s) close to the base will effectively jam the signals from the mobiles that are far away from the base. As illustrated by Figure 3.16, the observed capacity of the CDMA system is closely related to the accuracy of power control.

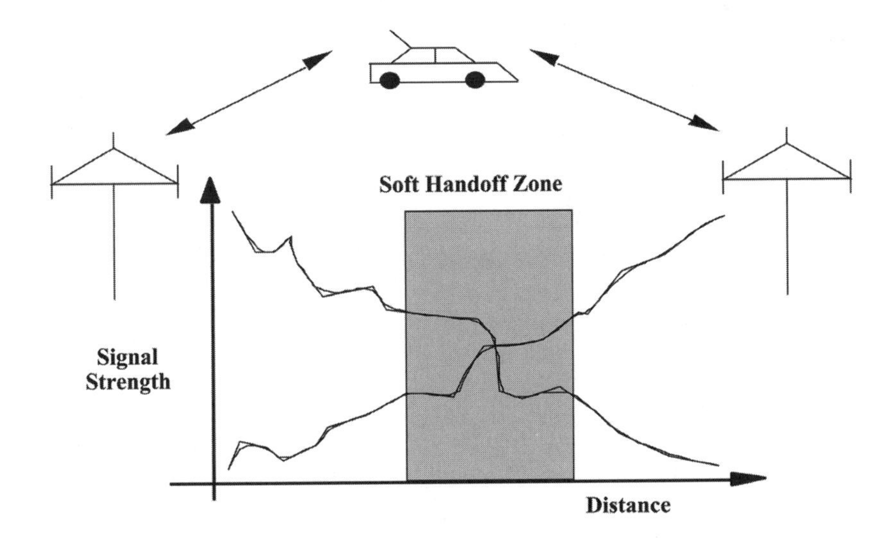

Figure 3.15. Illustration of *soft handoff* in the CDMA cellular system.

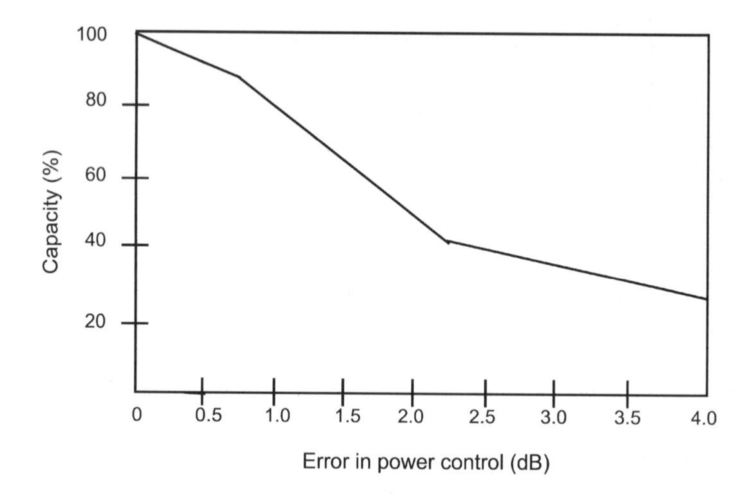

Figure 3.16. Sensitivity of CDMA system capacity to errors in power control.

Thus, for the CDMA system to work effectively, RF power in the system needs to be controlled. These power control requirements have the following two key components:

- All the transmissions from the mobiles must be received at the base station's receiver at approximately the same strength (within 1 dB), even under fast multipath fading conditions.
- In order to maximize the number of users sharing a cell, only the minimum RF power required for reliable communication should be allowed from the base station transmitter.

To meet the first requirement, CDMA systems deploy open-loop and closed-loop controls. The open loop control is used to counter the effect of wide dynamic range of the path loss (between the farthest and the closest mobile to the base station), which is on the order of 80 dB. In the open-loop control, the mobile station estimates the loss between itself and the base station and then uses it to make a coarse adjustment in its transmitted RF power. The path loss estimate is carried out by the mobile by measuring the received signal level (from the base station). The mobile then combines this reading with the power control information sent by the base station during some initial signaling transactions. This information includes the transmit power from the base station and some parameters that can be used to adjust the open-loop power control for different-sized cells and different cell ERP (effective radiated power) and receiver sensitivities. Whereas the open-loop power control attempts to provide a coarse adjustment to RF transmit power from a mobile, the closed-loop control actually measures the SNR at the cell-site receiver and, based on a predetermined threshold (desired SNR), instructs the mobile to adjust its transmit power. In practice, the cell receiver forms an estimate of the received SNR of the mobile's signal every 1.25 ms. The SNR measurement is compared with the set point or threshold value of the SNR. If the received SNR is too high/low, a decrease/increase power command is sent to the mobile. This power control command is sent every 1.25 ms, providing an 800-b/s power control rate. The mobile responds to the power control bits ($1 \Longrightarrow$ decrease power and $0 \Longrightarrow$ increase power) by making small adjustments of 2.0 dB in its RF power output until the received SNR at the cell site is within acceptable limits (within 1.0 dB of the desired set point).

To meet the second power control requirement—that is, to minimize the base station transmit power and maximize the number of mobiles that may share a cell—the base station continuously lowers its transmit power in small steps until the responding mobile signals the base station for more power. This process is performed individually for each of the CDMA channels on a forward CDMA waveform.

Thus the CDMA system capacity is very sensitive to the accuracy of power control. In a CDMA cellular mobile system the accuracy of power control is limited by the continuous movement of mobile stations, and it is considered that power control accuracy within 1 dB is the best that can be achieved. However, in a CDMA-based wireless local loop system with no mobility, a very high level of accuracy in power control can be achieved because the terminal locations are fixed. Significant gain in system capacity can therefore be achieved (compared to mobile systems) in CDMA-based WLL systems. For CDMA systems with limited mobility, there will be a loss in capacity because of increased errors in power control (and also because some channels may have to be reserved in each cell/sector for handoffs).

3.6.4.4 *Synchronization (Timing Control)* Use of code sequences to distinguish cells and sectors require stringent timing control (synchronization) among base stations. Timing control is also required between mobiles (terminals) and serving base station. Synchronization among base stations can be achieved by using a master clock with frequent broadcast of timing signals. However, if the base stations are scattered far apart, (leading to unacceptable propagation delays), it may be necessary to deploy the assistance of the GPS (global positioning system) for timing control.

Synchronization of subscriber terminals and base stations (up-link) is almost impossible to achieve in a cellular mobile system because of the movement of the mobiles. The best that can be achieved is through the use of the PN codes, which can be reproduced at the BS, though they result in a certain level of interference. However, in a CDMA WLL system with fixed terminals or with very limited mobility, better synchronization can be achieved through the use of time advance signals from the serving base station at the time of terminal setup.

3.6.4.5 *Soft Capacity* The CDMA radios based on the spread spectrum concept are designed to tolerate some level of interference, with their overall capacity limited by how well this mutual interference can be controlled. The capacity of conventional radios based on FDMA or TDMA concepts is limited by the number of noninterfering signals achieved by using complete separation either in frequency or time. Thus the capacity of CDMA systems is interference-limited, while the capacity of FDMA or TDMA systems is limited by the number of noninterfering coordinates in terms of frequency bands and/or time slots.

In the present U.S. cellular environment, each operator can deploy 12.5 MHz of spectrum, which provides 57 analog channels in a three-sector cell site. When demand for service is at a peak, the 58th request for a channel in the cell must be blocked and given an all-channels busy signal. With the CDMA systems, however, there is a much softer relationship between the number of users and the (transmission) grade of service. In other words, an operator has the flexibility to admit additional users during peak periods by providing some what degraded grade of service (increased bit error rates). This capability is especially important for avoiding dropped calls during handoff because of a lack of available free channels.

3.7 DECT: EUROPEAN DIGITAL CORDLESS TELECOMMUNICATIONS STANDARD

3.7.1 Introduction

Analog cordless telephones are in common use in residential applications where the telephone cord is replaced by a wireless link to provide terminal mobility to the user within a limited radio coverage area. Digital cordless telecommunication systems are intended to provide terminal mobility in residential, business, and public access applications where the users can originate and receive calls on their portable terminals as they change locations and move about at pedestrian speeds within the coverage area. It is also anticipated that the same terminal can be used in the three application environments—that is, at the residence, at the work place, and at such public locations as airports, train and bus stations, shopping centers, and so on.

In contrast to cellular radio, cordless telecommunications standards primarily offer an access technology rather than fully specified radio access and network standards. Cordless terminals generally transmit at lower power than cellular, resulting in the use of microcells. In high-density (in building) applications, much smaller cells (pico cells) may be used so that significantly higher traffic densities can be supported. Furthermore, cellular networks generally operate in an environment characterized by regulation (which may vary from country to country), limited number of operators, centrally managed frequency resources, and relatively high infrastructure costs. Cordless telecommunication systems, on the other hand, are expected to operate in an unregulated, open market environment where system installation and (frequency) planning cannot be coordinated or planned, and the end-user perceived cost and performance are key factors in market acceptance.

The European digital enhanced cordless telecommunications (DECT) standard was developed by ETSI to serve a range of applications that included

- Residential systems
- Small business systems (single site, single cell)
- Large business systems (multisite, multicell)

- Public cordless access systems
- Wireless data (LAN) access systems

The DECT standard has a modular structure so that requirements of different applications can be accommodated in a cost-effective manner. For example, data LAN and PABX applications represent closed environments requiring minimum open standards, whereas with an application like public cordless access, it is essential that products from different vendors not only coexist but interwork with each other. The DECT standard therefore includes different (so-called) *profiles*, each of which represents a set of functionalities that a customer can choose to implement to support such applications as public cordless access, PABX, LAN, Data, GSM/DECT interworking, and so on.

3.7.2 Functional Architecture for DECT System

The DECT standard can support high local user density and is suitable for operation in unpredictable and highly variable traffic and propagation conditions. Though DECT is primarily a radio interface specification, the DECT standard also includes a network architecture without specifying all the necessary procedures and protocols. A functional or conceptual view for a DECT system is shown in Figure 3.17.

Independent of the application, a DECT system is composed of a number of functional entities that include

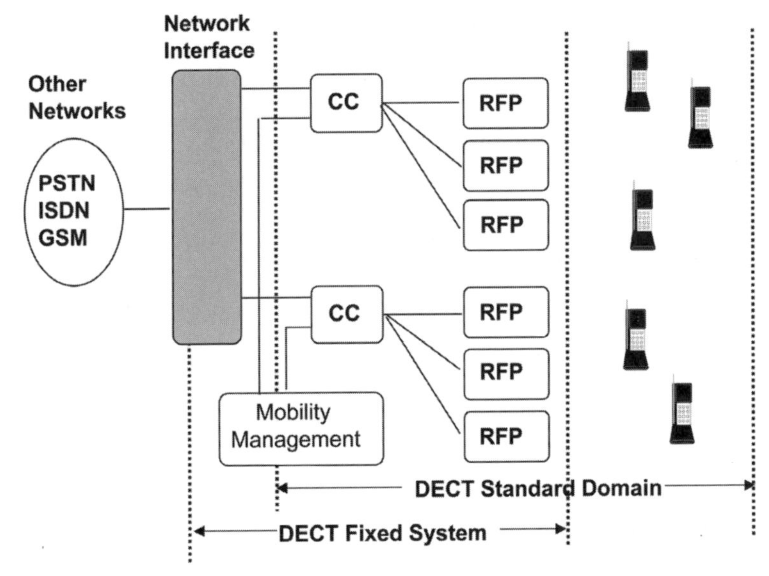

Figure 3.17. Functional view of the DECT system.

- A DECT portable handset or terminal that may include a cordless terminal adapter (CTA) to support more general applications (e.g., fax, data)
- A DECT radio fixed part (RFP) that supports the physical layer of the DECT common air interface
- A cordless cluster controller (CCC) that supports the MAC, DLC, and network layers for one or more clusters of RFPs and that, in a multicell environment, handles intercell handoffs
- A network-specific interface unit (IU) or interworking unit that provides required interfaces to connect to such specific networks as PSTN, ISDN, GSM, and so on, and, in a multicell environment, may also provide call/connection control for DECT terminals
- A mobility management function that supports centralized authentication for public access

3.7.3 Radio Aspects for DECT System

In Europe the 1880- to 1900-MHz band has been set aside for DECT. In order to utilize the available 20-MHz band in an efficient and flexible manner for supporting voice and data applications, the DECT standard provides for space, frequency, and time distribution. Space dispersion in DECT is supported through the frequency reuse feature based on cellular concept. To provide frequency distribution, the available spectrum is segmented into 10 carrier frequencies (frequency channels) from 1881.792 to 1897.344 MHz with a separation of 1.728 MHz—that is, by deploying frequency division multiple access (FDMA). Time distribution is achieved by using time division multiple access (TDMA) and time division duplex (TDD) methods whereby each frequency channel supports 12 duplex time slots or 32-kb/s channels.

The FDMA/TDMA/TDD channel structure for the DECT system was illustrated in Figure 2.5. The key radio parameters associated with DECT are summarized in Table 3.4.

TABLE 3.4. Typical Radio Parameters Associated with the DECT System

Radio Parameter	DECT Value
Multiple access/duplex method	TDMA/TDD
Spectrum allocation	1880–1900 MHz
Carrier spacing	1728 kHz
Number of carriers	10
Channels/carrier	12
Modulation method	GFSK
Transmission rate	1152 kb/s
Speech coding	32-kb/s ADPCM
Frame duration	10 ms
Peak output power	250 mW

A total of 120 channels, each with 32-kb/s data rate, are available in this FDMA/TDMA/TDD structure to carry voice and data traffic, with additional capacity required for business systems being provided by invoking frequency reuse in a cellular configuration. In the cellular configuration used in business systems and in low-mobility public systems, DECT also supports handoffs across cells.

Though the basic data rate for DECT is 32 kb/s, a number of time slots (not necessarily adjacent or in the same frequency channel) can be concatenated to provide higher-bit-rate channels; for example, 64 kb/s using two time slots or 384 kb/s using 12 time slots are defined in DECT for ISDN and LAN applications, respectively. Higher-capacity channels can be set up in both the simplex and duplex modes. DECT standard also defines half-rate channels (using half-time slots) that can be used for deployment of low-bit-rate coders.

The mechanism for channel selection in DECT is known as continuous dynamic channel selection (CDCS) whereby the channel selection is done on the basis of least interference within all the available DECT channels (any time slot on any frequency carrier). When a connection is needed, the portable selects a channel with least interference (highest quality) and continues to scan the remaining channels. If, during the call, a better channel becomes available, the portable switches to this channel. The old and the new connections can overlap in time so that a seamless handoff can take place.

3.7.4 DECT Network Aspects

DECT supports a diverse set of services and applications which requires a clear distinction between the processes that are part of the DECT specification and those that the DECT network simply supports. This, in turn, requires a structured protocol architecture based on the OSI principles. The DECT architecture follows the OSI layering and is shown in Figure 3.18.

A primary object of the DECT standard has been to ensure maximum applicability of the standard in a variety of environments based on existing or emerging public and private networks. Thus, the DECT radio interface and the related radio link protocol architecture has been defined so that DECT users can access a range of networks, particularly

- Analog PSTN, PBXs, and key systems
- Digital ISDN public networks, PBXs, and key systems
- GSM networks
- Local area networks
- Public data and IP networks

As shown in Figure 3.18, the flexibility in the DECT system is provided by the network interface unit that allows the DECT terminals to connect to a range of diverse public and private networks. The radio link protocol architecture provides the necessary hooks to facilitate such interworking with external networks (at the

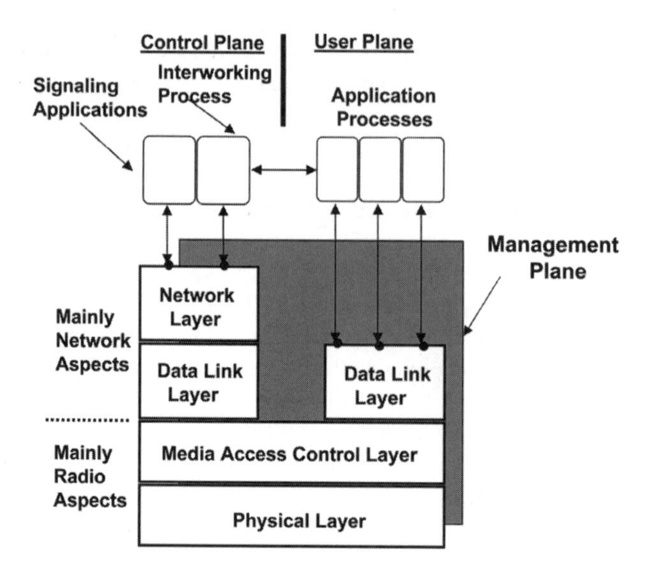

Figure 3.18. Layered architecture for the DECT system.

control level as well as at the user information level) in the form of service access points (SAPs) to signaling applications, interworking processes, and application processes.

3.8 PHS: JAPANESE DIGITAL CORDLESS TELECOMMUNICATIONS STANDARD

The Japanese PHS (personal handy-phone system) standard for cordless telecommunications was completed at the end of 1993 by the Research and Development Center for Radio Systems, which is now known as the Association of Radio Industries Businesses (ARIB) of Japan. The PHS addresses similar application environments as DECT (e.g., residential, business and public cordless access) using a single low-power terminal. The PHS public access service was introduced in Japan in 1994. Besides the basic telephony service, PHS also provides data services including a 32-kb/s data service and has been used as the underlying technology for WLL systems. Some features and capabilities associated with PHS include

- Error detection using CRC (PHS does not provide error correction)
- Dedicated control channels
- Dynamic channel assignment
- Handoff (as an option) at walking speeds
- Transmission diversity on the forward link (base to handset)
- Group 3 fax service at 2.4–4.8 kb/s

- Full duplex modem transmission at 2.4–9.6 kb/s and at 32 kb/s
- Terminals with standby power for up to 800 hours

Besides the application of PHS in the public (cordless) environment, PHS terminals can also be used in a business (wireless PBX) and residential cordless environment. In fact, one of the distinct features of the PHS system is that a single terminal can be used in all three application environments. These application environments are illustrated in Figure 3.19 along with the type of interfaces required for interconnecting to the public ISDN network.

3.8.1 PHS Radio Aspects

The spectrum allocation for PHS is in the 1895- to 1918.1-MHz band and is partitioned into 77 carrier frequencies with a separation of 300 kHz. Each carrier is then time division-multiplexed into two groups of four time slots operating in a time division duplex (TDD) mode to provide four duplex channels per carrier. Forty of the 77 carriers (1906.1–1918.1 MHz) are allocated for public systems, and the remaining 37 (1895–1906.1 MHz) are allocated to home/office applications. As in the DECT system, dynamic channel assignment is deployed by PHS whereby the channel selection is autonomous based on measured signal strength. Key radio parameters for PHS are summarized in Table 3.5. The signal represented by each carrier is time division-multiplexed into eight time slots; four for down-link and four for up-link transmissions, to provide four full duplex channels in time division duplex (TDD) operation as shown in Figure 3.20.

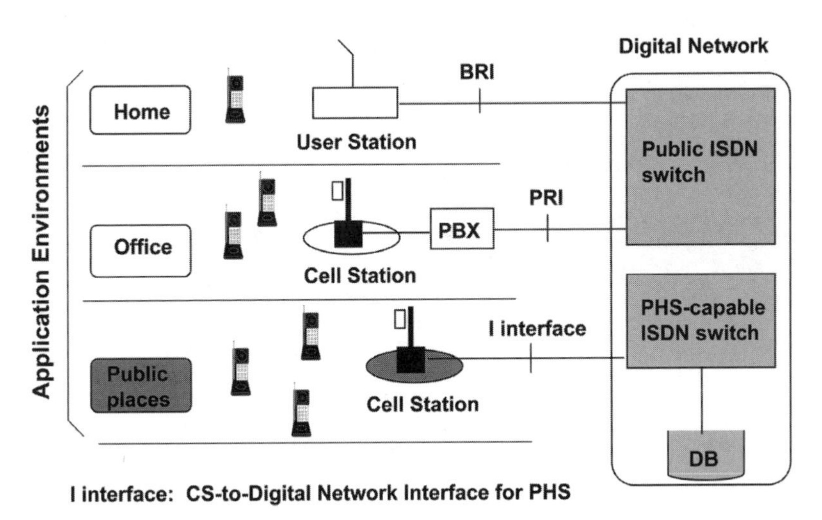

Figure 3.19. Different application environments for the PHS system.

TABLE 3.5. Typical Radio Parameters Associated with the PHS System

Radio Parameter	PHS Value
Multiple access/duplex method	TDMA/TDD
Spectrum allocation	1895–1918 MHz
Carrier spacing	300 kHz
Number of carriers	77
Channels/carrier	4
Modulation method	Pi/4shifted QPSK
Transmission rate	384 kb/s
Speech coding	32-kb/s ADPCM
Frame duration	5 ms
Peak output power	80 mW

3.8.2 PHS Network and Protocol Aspects

The general requirements for providing uninterrupted and consistent service to moving PHS stations (PSs) include

- Providing subscribed services to a PHS user regardless of PS location
- Charging to the same PS based on the distance between the called and calling parties depending on PS location
- Maintaining location information for all the PS in the system

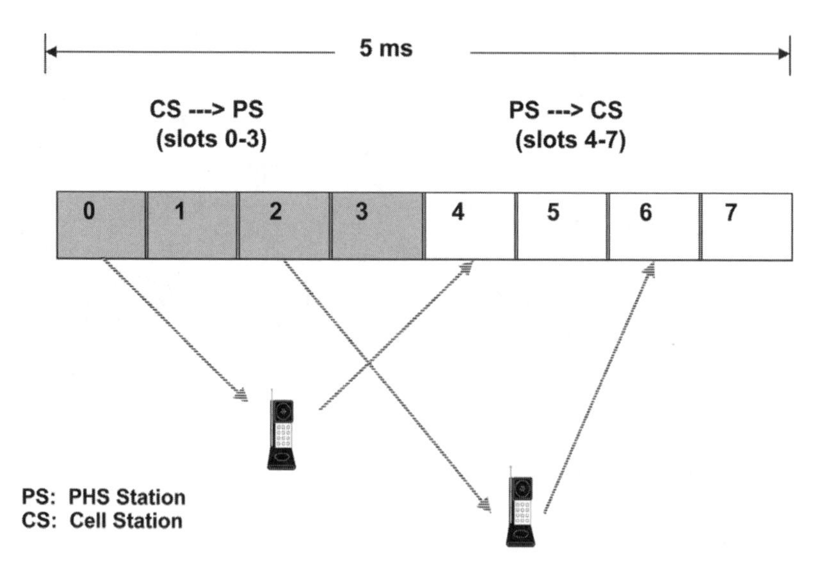

Figure 3.20. TDMA/TDD time-slot arrangement for the PHS system.

- Routing and delivery of incoming calls to the current location of the PS
- Authentication of the PS as required by the PHS service provider

Support of these requirements is greatly facilitated by the signaling capabilities available in digital networks and by the intelligent network (IN) capabilities that are rapidly being introduced. Some of the capabilities required from the signaling protocol for the I interface between the PHS cell station (CS) and the network include

- Handling of both voice and data communications
- Support of mobility features (registration, authentication, incoming call delivery)
- Maintaining independence between CS and the digital network
- Support of required supplementary services
- Future proof against addition of new services and functions
- Flexibility to support multiple PHS operators

For PHS system implementation, it is possible to deploy stand-alone PHS networks where all the mobility management functions (like location registration and authentication) and required databases are provided by the PHS service operator. Some PHS operators in Japan have opted for this approach mainly for security reasons. An alternate, and a more commonly used deployment scenario, is to utilize the capabilities available in the local ISDN network. In this configuration, all of the network functions for PHS including location registration and authentication are provided by the local ISDN operator, and only cell stations (CSs) are provided by the PHS service operator. This type of configuration is not only more economical but also speeds up the introduction of new PHS service offerings. The PHS operators using this option pay a call-by-call-based access charge to the local ISDN operator. The charge includes the cost of provision and use of the location registration and authentication database as well as the cost of local exchange ports that support the required CS-to-ISDN interface.

In order to support the latter scenario, a standardized interface for interconnecting the PHS cell stations to the public ISDN network is required. The relevant reference points for interconnection of a CS to the digital network are shown in Figure 3.21. The PHS service provider has the option to support the CS-to-digital network interface at reference point X1 or X2. In the former case the network termination is considered to be a part of the digital network and provides some advantages from a network management point of view and requires no changes to the interface if the transmission medium is upgraded (e.g., from copper to fiber). For the X2 reference point the NT is considered to be a part of the CS, and an interface defined at this reference point is better economically because internal (proprietary) signaling can be used in the CS–NT link before converting to a standardized signal at the X2 reference point. The I interface for the

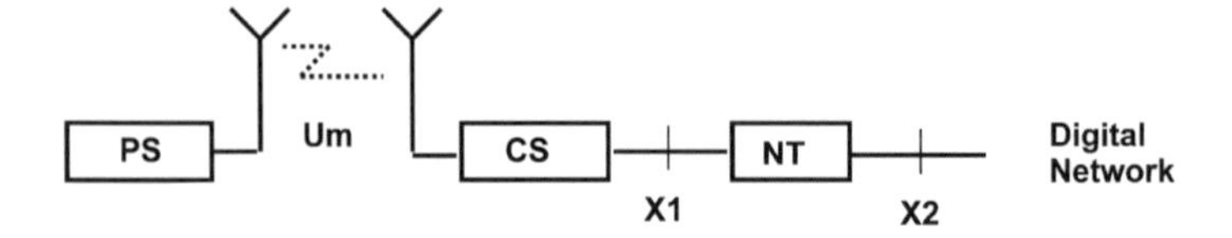

Figure 3.21. Interface reference points for PHS access network.

interconnection of PHS cell stations to the ISDN network is based on the following ITU-T Recommendations:

- Layer 3: Q.931 and Q.932
- Layer 2: Q.920 and Q.921
- Layer 1: I.431 (X1 reference point) and G.961 (X2 reference point)

3.9 MOBILE SYSTEM ENHANCEMENTS AND THIRD GENERATION (3G) STANDARDS

The brief description of cellular mobile systems provided in the preceding sections reflects the basic capabilities envisaged in the initial standards for these systems which essentially focused on circuit-switched voice and voice-band data services. The initial development and commercial deployment of these digital cellular systems took place in the early 1990s. Since then these systems, sometimes referred as second-generation (2G) mobile systems, have been undergoing enhancements and improvements in terms of their technical and service capabilities. With the rapid increase in demand for high-bit-rate data and Internet services, major efforts have been made to evolve the various cellular mobile technologies to efficiently accommodate the emerging needs for high-bit-rate data services, especially packet-switched data services. TDMA- and CDMA-based mobile communication systems are now available that can support data rates over 128 kb/s and a range of services including multimedia message service (MMS), e-mail, other Internet services, and location-based services.

Besides the natural evolution of 2G mobile systems to accommodate the user demand for enhanced voice and data services, standards for new radio technologies and network architectures have also been developed that define third-generation (3G) mobile communication systems. These standards and specifications have been developed by various regional and international standardization organizations under the ITU (International Telecommunications Union) standard known as IMT-2000 (International Mobile Telecommunications—2000). ITU has identified radio spectrum in the 2000-MHz band for IMT-2000, and these systems are starting to be deployed in many countries.

3.9.1 Enhancements to TDMA Cellular Systems for Higher Data Rates

A significant number of services and service capabilities were progressively added to the initial GSM and IS-136 standards as part of subsequent phases or versions of these standards. Enhancements for voice services generally address higher traffic/call handling capabilities as well as provision of a wide range of intelligent network (IN) and ISDN services. A major effort is devoted toward increasing data handling capabilities so that emerging demand for high-bit-rate data and Internet services can be accommodated. This section therefore focuses on the enhancements related to higher-bit-rate data capabilities of these systems. Although these increased data rate capabilities were originally developed for the GSM in Europe (and the description below reflects this), they have also been incorporated in the IS-136 TDMA system in North America.

3.9.1.1 High-Speed Circuit-Switched Data (HSCSD) This is an extension of the initial circuit-switched data services in GSM which supported only 9.6-kb/s voice-band data services. The basic approach in HSCSD is to achieve higher user bit rates by using multiple TDMA time slots for a data connection while keeping the GSM physical layer implementation for data services unchanged. The use of multiple time slots within a GSM frame structure of eight time slots per 200-kHz carrier leads to achievable user rates of up to 76.8 kb/s (9.6×8 kb/s). For simplifying the design of the mobile station and for minimizing changes to the current architecture, multiplexed time slots are chosen to be consecutive time slots. Besides providing higher data rates, HSCSD also provides the potential for *pseudo-asymmetric* operation whereby some of the subchannels assigned for the up-link direction are not used, resulting in a larger capacity for the down-link transmissions—an option suitable for Internet applications.

Most of the functions for data services in GSM are located in the mobile station (terminal adaption) and the mobile switching center which provides the necessary interworking functions (IWF) like rate adaption, error control, and modems. The HSCSD utilizes the same implementation by treating the multiple channels assigned to a HSCSD call as independent subchannels as far as the rest of the GSM network is concerned. The channel splitting and combining functions need to be implemented only at the two ends (i.e., at the MS and the MSC, respectively). The HSCSD architecture is illustrated in Figure 3.22.

Though HSCSD is based on the apparently simple principle of multiplexing up to eight time slots without requiring major changes to the GSM network infrastructure, full advantage of HSCSD in terms of highest bit rates can only be achieved by implementing some additional changes to existing protocols and procedures. Examples of such changes include: a new framing protocol for the BSS-to-MSC

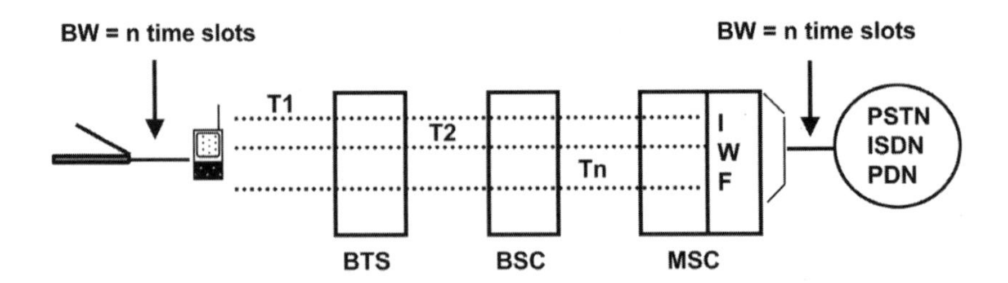

Figure 3.22. Architecture for high-speed circuit-switched data (HSCSD).

link, modification to frequency hopping procedures, extensions to the radio link protocol (RLP), and signaling procedures to support flexible bearers. Many of these complexities can be avoided, and the system design can be simplified if the HSCSD implementation is restricted to lower data rates (e.g., 28.8 kb/s) where multiplexing of only two or three subchannels is sufficient.

3.9.1.2 General Packet Radio Service (GPRS) This is intended for the data environment where a significant portion of emerging mobile data communications applications are based on bursty Internet applications like e-mail and *world wide web (www)* browsing for which use of a circuit-switched connection is wasteful and expensive. In order to capture the market potential of such data applications, GSM introduced the general packet radio service (GPRS). GPRS maintains the core GSM radio access technology and provides packet data services by introducing two new network elements called serving GPRS support node (SGSN) and Gateway GPRS support node (GGSN). In addition, the GPRS register, which may be integrated with the GSM HLR, maintains the GPRS subscriber data and routing information. GPRS will operate at transmission data rates from 14.4 kb/s to 115.2 kb/s by using from one up to eight time slots in GSMs TDMA frame structure. The network architecture for GPRS is illustrated in Figure 3.23. The architecture attempts to ensure that little or no hardware modifications are required for the existing GSM network elements and that same transmission links between the BTS and the BS can be utilized. Furthermore, GPRS will utilize the existing GSM authentication and privacy procedures, and the GPRS register which contains the GPRS subscriber data and routing information can be integrated with the HLR. Besides the GPRS register, two new nodes are defined for GPRS.

The serving GPRS support node (SGSN) is responsible for communication between the mobile station (MS) and the GPRS network. The primary functions of the SGSN are to detect and register new GPRS mobile stations in its serving area, send/receive data packets to/from the GPRS MS, and track the location of the MS within its service area. The GGSN acts as the GPRS gateway to the external networks and provides such interworking functions as translation of data formats, protocol conversions, and address translation for routing of outgoing and incoming packets.

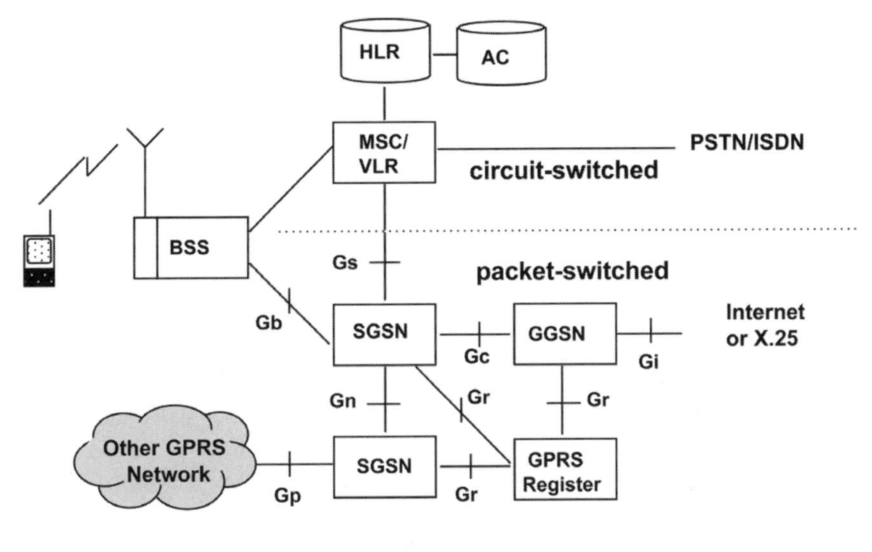

Figure 3.23. Network architecture for general packet radio service (GPRS).

3.9.1.3 Enhanced Data for Global Evolution (EDGE)

As indicated in the preceding two sections, HSCSD and GPRS essentially represent add-on capabilities to the basic voice-optimized TDMA cellular networks, without any significant changes to the radio interface technology. Any further increases in user data rates are constrained by the interference-prone nature of the radio environment and can only be achieved by suitably modifying the radio access part. The EDGE concept proposes to use suitable combinations of modulation schemes and channel encoding rates as part of a *link adaptation* scheme. Thus, in order to optimize the data throughput over the radio channel, the combination of the modulation method and the channel encoding rate is selected adaptively based on the prevailing interference levels on individual channels (i.e., based on the time-varying link quality).

One of the fundamental characteristics of a cellular mobile system is that different users encounter different channel qualities in terms of signal-to-interference ratios (SIRs) as a result of different distances to the base station, different levels of fading, and different interference environments. These differences in channel quality for individual users are mitigated to some degree by implementing power control. Typically, the cellular mobile system is configured to ensure that only a very small percentage of users encounter a SIR below the threshold and thus receive a reduced or unacceptable quality of service. The SIR for the remainder (majority) of the users, on the other hand, is above the design target, and they receive quality of service that is good to excellent and is essentially independent of the channel quality. In other words, under this scenario a large percentage of user population will encounter an excellent channel quality from which they cannot derive any benefit because the radio interface is unable to adjust its throughput to reflect the prevailing better than average channel quality.

The data throughput for a given channel bandwidth (or spectrum efficiency) is determined by (a) the information bit rate (baud rate) supported by the modulation method and (b) the level of redundancy in the channel coding scheme used by the radio interface technology. The higher the baud rate associated with a modulation method and the lower the level of redundancy in the encoding scheme, the higher the raw data rate across the radio channel. However, higher-level modulation methods imply higher complexity in radio design, and generally a modulation method like 8-PSK (phase shift keying) provides a good compromise. On the other hand, in order to maintain an acceptable level of protection from channel errors, the level of encoding redundancy can be adjusted according to the channel quality so that when the channel is less error-prone, encoding redundancy can be reduced, thereby increasing the user data rate over the radio channel. This principle is utilized for the GSM/EDGE radio interface where both GMSK and 8-PSK modulation methods and various encoding redundancy levels are applied to individual radio channels based on its quality. Different code rates (redundancy levels) are generated by puncturing a different number of bits from a common convolutional code (ratio 1/3).

Technically, EDGE is primarily a TDMA radio interface improvement, but in a more general context it can also be viewed as a systems concept that allows the GSM and IS-136 TDMA networks to offer a set of new bearer services including such packet-switched and circuit-switched bearers as GPRS and HSCSD, respectively. EDGE also provides an evolutionary path for GSM and IS-136 TDMA systems for delivering 3G mobile services in existing cellular mobile radio-frequency bands. Specifications for GSM/EDGE are being developed by the European Telecommunications Standards Institute (ETSI) as part of the GERAN (GSM/EDGE Radio Access Network) project. EDGE-based radio interface requires a base station system with necessary modifications to accommodate GPRS and/or HSCSD services, and these systems can provide user data rates to 384 kb/s.

3.9.2 Enhancements to CDMA Cellular Systems for Higher Data Rates

The initial CDMA specification contained in the TIA IS-95 standard has undergone further enhancements leading to versions IS-95A and IS-95B. Enhancements to an IS-95 CDMA system after revision IS-95B represent development of the system to support 3G mobile services and capabilities. The progress of IS-95 CDMA evolution is summarized in Table 3.6.

CDMA 1XEV-DO (sometimes also referred as CDMA 1XEV phase 1) represents the most significant enhancement in terms of supporting high-bit-rate data and Internet services for this CDMA technology. Optimized for packet data services, CDMA2000 1XEV-DO provides a peak data rate of 2.4 Mbps within one 1.25-MHz CDMA carrier. It leverages the existing suite of Internet protocols (IPs), and hence it supports all popular operating systems and software applications. 1XEV-DO offers

TABLE 3.6. Evolution of an IS-95 CDMA System to Support Third-Generation (3G) Services

CDMA System	Capabilities	Remarks
IS-95A	Voice + 14.4-kb/s data	2G CDMA system first deployed in 1996 (TIA standard)
IS-95B	Voice + 64-kb/s packet data	2.5G CDMA system first deployed in 1999 (TIA standard)
CdmaOne	IS-95A + IS-95B + ANSI-0008	Family of services including PCS & WLL (cdmaOne is a trademark of CDG)
CDMA 1X	2 × 2G voice + 144-kb/s data	3G CDMA system first deployed in 2000 (ITU 3G standard)
CDMA2000 1XEV-DO	2.4 Mb/s (a separate 1.25-MHz carrier is used for data only)	ITU 3G standard
CDMA2000 1XEV-DV	Voice + 2- to 5-Mb/s data using 1.25-MHz carrier	ITU 3G standard pending
CDMA2000 3X	3G voice + data services using 3 × 1.25-MHz carrier	ITU 3G standard

CDMA 1X: uses one 1.25-MHz carrier.
CDMA 3X: will use 3 × 1.25 MHz = 3.75-MHz carrier.
CDMA 1XEV-DO: CDMA 1X evolution—data only.
CDMA 1XEV-DV: CDMA 1X evolution—data voice.

an *always-on* user experience, so that users are free to send and receive information from the Internet and their corporate intranets—any time, anywhere.

Due to the highly asymmetric nature of most data services, the data rates on the forward channel or down-link (base station to subscriber terminal) are critical and need to be maximized. In the original IS-95 forward channel, a multitude of low-data-rate channels are multiplexed together (with individual channels separation ensured by assigning different *codes* to them) and share the available base station power with some form of power control. This is an optimal option when multiple low-rate users are to share common bandwidth, but not when the same bandwidth is to be shared by a few high-rate data users. The scheme becomes even more inefficient when the same bandwidth is shared between low-rate voice and high-rate data users. The solution proposed by CDMA 1XEV-DO is to use, possibly adjacent but nonoverlapping, bandwidth allocations for low-rate voice (and voice-band data) and high-data-rate (HDR) applications.

With a dedicated RF carrier for HDR users, the forward link for 1XEV-DO takes on a different form from the regular IS-95 CDMA system. In the 1XEV-DO the down-link transmissions are time-multiplexed and are transmitted at full available base station power, but with data rates and time slot sizes determined by the quality of the user channel. During idle periods (no user packets queued up), the channel is silent except for short pilot and control packets, thereby eliminating potential interference to adjacent channels.

The short pilot bursts sent by the base stations on the forward channel are used by the subscriber terminal to estimate the best signal-to-interference ratio (SIR) on the channel, which is then mapped to a value representing the maximum data rate that can be supported on the channel. The mapping of the SIR on the channel to the data rate corresponds to one of the values indicated in Table 3.7 which also indicates the optimal packet size, modulation method, and channel coding rate associated with the selected data rate. The high-level modulation methods include 8-PSK and 16-QAM, and the FEC is based on turbo coding which is appropriately punctured to achieve higher data rates.

The information on the maximum data rate for the down-link transmission is communicated to the appropriate base station on the reverse or up-link channel using the *data request channel* (DRC), and this information is updated as frequently as every 1.67 seconds. Based on the information on the DRC, the base station starts the transmission of user data at the requested data rate, thus ensuring maximum channel data throughput in real time.

The basic network architecture for deployment of CDMA 1XEV-DO is illustrated in Figure 3.24. It consists of a multicarrier base transceiver station, an integrated base station controller (BSC), a packet control function (PCF), and a packet data service node (PDSN). 1XEV-DO system does not utilize any mobile switching center (MSC) functions. Each BTS includes both RF and digital components for multiple 1.25-MHz RF carriers at the cell site. The BTS terminates the radio link and provides radio link protocol (RLP) and universal data link (UDL)

TABLE 3.7. Various Data Rates and Associated Parameters in CDMA 1XEV-DO

Data Rate (kb/s)	Packet Size (bytes)	Turbo Code Rate (b/s)	Modulation
38.4	128	1/5	QPSK
76.8	128	1/5	QPSK
102.6	128	1/5	QPSK
153.6	128	1/5	QPSK
204.8	128	1/3	QPSK
307.2	128	1/3	QPSK
614.4	128	1/3	QPSK
921.6	192	1/3	QPSK
1228.8	256	1/3	QPSK
1228.8	256	1/3	16-QAM
1843.2	384	1/23	8-PSK
2457.6	512	1/3	16-QAM

Source: E. Esteves, The high data rate evolution of the cdma2000 cellular system, in *Multiaccess, Mobility and Teletraffic for Wireless Communications*, Vol. 5, Kluwer Academic Publishers, Hingham, MA (2000), pp. 61–72.

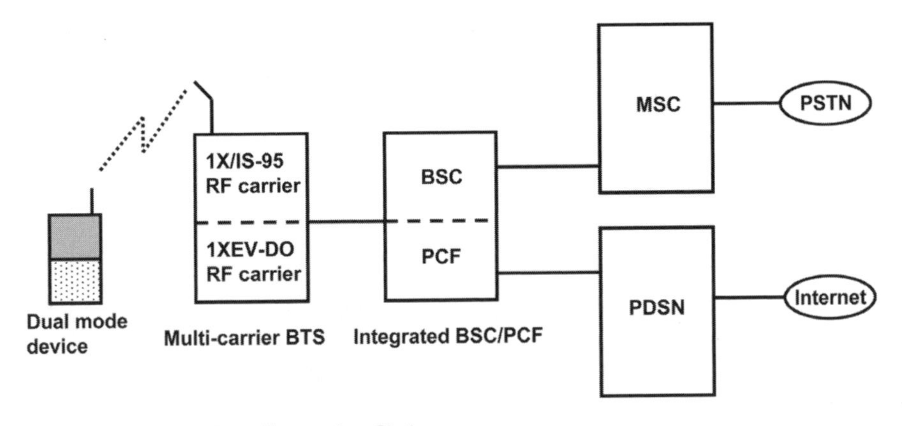

BTS: Base Transceiver Station
BSC: Base Station Controller
PCF: Packet Control Function
MSC: Mobile Switching Center
PDSN: Packet Data Service Node

Figure 3.24. Basic architecture for CDMA 1XEV-DO system deployment.

connectivity to the BSC using packetized backhaul. Each 1XEV-DO sector can provide peak data rates of 2.4 Mb/s on the forward link and 153.6 Kb/s on the reverse link.

The BSC performs session control, connection control, and selection functions to support each user's session. The PCF is integrated into the BSC and provides additional functionality to handle packet data. It manages the necessary routing interface to the PDSN, selection of appropriate PDSN, and tracking of idle devices. The PDSN performs traditional network access server (NAS) functions like point-to-point protocol (PPP) support. An off-the-shelf remote authentication dial-in user service (RADIUS) server is used for authentication, authorization, and accounting (AAA) functions. Other off-the-shelf components like DNS (domain name system) and DHCP (dynamic host configuration protocol) are used to support Internet connectivity and service support.

3.9.3 Emerging Third-Generation (3G) Mobile Communication Systems

As mentioned earlier, the most of the digital cellular and cordless radio currently deployed around the world represent the so-called second-generation (2G) mobile and personal communication systems developed in the 1990s. In the meantime, work has been underway in the regional and international standards organizations to specify new and evolved radio interface and network standards to meet the needs of the new or third-generation (3G) mobile communications. These 3G systems are a part of a family of standards designated as IMT-2000 by the ITU.

The frequency spectrum between 400 MHz and 3 GHz is technically suitable for the third-generation mobile telecommunication systems. However, ITU-R has identified the following frequency bands for IMT-2000:

Up-link: 1885–2025 MHz
Down-link: 2110–2200 MHz

The frequency bands 806–960 MHz, 1710–1885 MHz, and 2500–2690 MHz, which are currently in use for cellular and microwave systems have also been identified for IMT-2000 applications (as they gradually become available in the future).

IMT-2000 offers the capability of providing value-added services and applications on the basis of a single standard. The system envisages a platform for distributing converged fixed, mobile, voice, data, Internet, and multimedia services. It is expected that IMT-2000 will provide higher transmission rates: (a) a minimum speed of 2 Mbit/s for stationary or walking users and (b) 348 kb/s in a moving vehicle. Second-generation systems only provide speeds ranging from 9.6 kb/s to 64 kb/s.

As a result of this work, ITU-R has developed Recommendation M.1457, which contains five terrestrial radio interface specifications for third-generation mobile

TABLE 3.8. Radio Transmission Technologies for Third-Generation Mobile Systems (IMT-2000)

IMT Radio Interface	Evolved From	Access Technology	Developed and Proposed By
IMT-DS (direct sequence)	New radio interface	Wideband CDMA (W-CDMA)	ETSI (Europe)
IMT-MC (multicarrier)	Evolved from 2G IS-95 (CDMA)	CDMA with multiple carriers (CDMA 2000)	TIA (N.A.)
IMT-SC (single carrier)	Evolved from 2G IS-136 (TDMA)	IS-136 with added features like GPRS, EDGE	TIA, UWCC (N.A.)
IMT-FT (frequency–time)	Evolved from 2G (DECT)	TDMA/TDD	ETSI (Europe)
IMT-TC (time code)	New radio interface	Combination of TDD and synchronous CDMA	ETSI (Europe) and CWTA (China)

systems that are generally referred as IMT-2000 radio transmission technologies (RTTs). These new radio standards for IMT-2000 are summarized in Table 3.8.

The current industry view is that out of the above five radio transmission technologies, the first two technologies (i.e., W-CDMA and CDMA 2000) are the front-runners in terms of availability of equipment and terminals and initial deployment of 3G systems. W-CDMA is a radio technology developed in Europe as the target 3G technology for evolution of current 2G GSM systems toward the universal mobile telecommunication system (UMTS)—the European 3G mobile communication system. The frequency assignment (FA) for W-CDMA is approximately 5 MHz, and it will use direct sequence CDMA (DS-CDMA) multiple access technology. The anticipated migration path from current GSM systems to W-CDMA-based UMTS is illustrated in Figure 3.25.

CDMA20003X radio transmission technology is considered to be an enhancement to and evolution from the IS-95A/B CDMA radio interface, and it represents the target 3G technology for evolution of currently deployed CDMA systems. It also utilizes direct sequence CDMA for multiple access, but the frequency assignments can be incremental in 1.25-MHz steps. Thus CDMA20001X is the initial version with a frequency assignment of 1.25 MHz which can ultimately be evolved into CDMA20003X with $3 \times 1.25 = 3.75$-MHz frequency assignment. However, the current view is that with the development and deployment of CDMA1XEV-DO and 1XEV-DV on the horizon, implementation of CDMA20003X systems may be far in the future. The anticipated evolution path

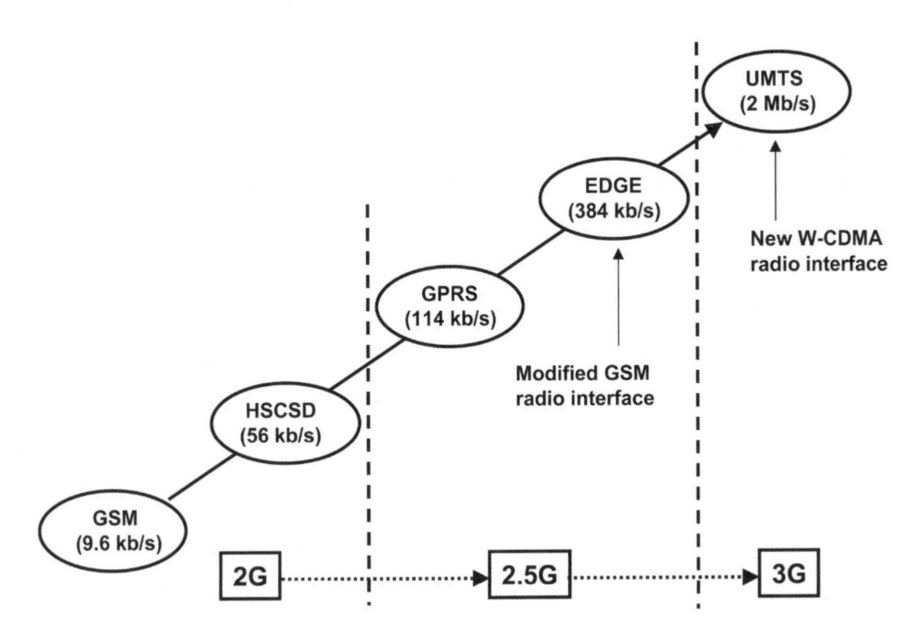

Figure 3.25. Evolution path from GSM (2G) to UMTS (3G) mobile systems.

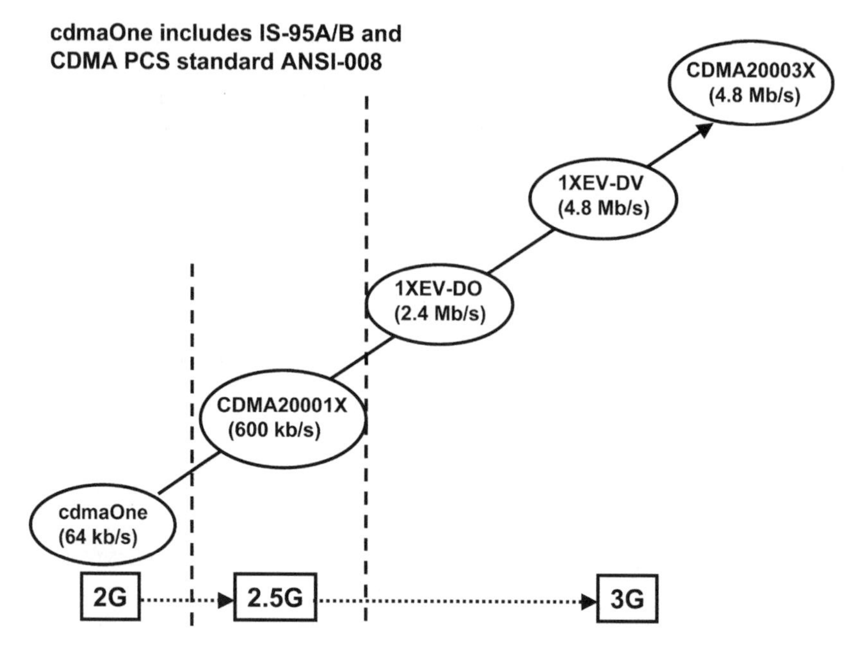

Figure 3.26. Evolution path from cdmaOne (2G) to CDMA2000 (3G) mobile systems.

from current (2G) CDMA systems to CDMA2000 (3G) mobile systems is illustrated in Figure 3.26.

The key radio parameters for W-CDMA and CDMA20003X radio interfaces are summarized in Table 3.9.

TABLE 3.9. Summary of Radio Parameters for W-CDMA and CDMA3X Radio Interfaces

Radio Parameter	W-CDMA System	CDMA20003X System
Multiple Access scheme	Wideband DS-CDMA	DS-CDMA
Duplex method	FDD	FDD
Channel spacing	4.4–5.0 MHz	1.25–3.75 MHz in steps of 1.25 MHz
Basic chip rate	4.096 Mcp/s	3.6864 Mcp/s (CDMA20003X)
Frame length	20 ms	10 ms
BS synchronization	Asynchronous	Synchronous
Forward Link pilot for channel estimation	CDM common	TDM dedicated
Antenna beam forming	CDM auxiliary	TDM dedicated
Channel coding	Convolutional and optionally RS codes	Convolutional and Turbo codes as necessary

In terms of the networks and protocols for these 3G systems, W-CDMA currently utilizes GSM-based network architectures with appropriately enhanced MAP protocols. CDMA2000 systems currently can operate with North American ANSI-41 networks as well as GSM-based networks and protocols.

Efforts are underway to develop specifications for *all IP* network architectures and protocols for 3G mobile systems so that these 3G radio technologies can seamlessly interoperate with each other and with the various components of the Internet. Services based on CDMA20001X and W-CDMA radio systems are now being offered in South Korea and Japan, and many other countries are on the verge of deploying 3G mobile communication systems.

CHAPTER 4

WLL System Components and Interfaces

As opposed to the fixed network, the subscriber equipment in a WLL system generally consists of complex electronic components to support the radio-related functions and ensure adequate service quality. Furthermore, a reliable power source for the subscriber terminal is essential for uninterrupted service including a standby power source. The radio interfaces available for WLL applications include those based on existing and emerging cellular and cordless telecommunication system radio standards, as well as those based on proprietary technologies. There are also a number of options for network interfaces for interconnecting the WLL radio access network to the PSTN switch. This chapter describes typical WLL system components and available options for radio and network interfaces.

4.1 WLL REFERENCE MODEL AND MAJOR SYSTEM COMPONENTS AND INTERFACES

A general reference model for a WLL system is shown in Figure 4.1. It illustrates the relevant functional elements (FEs) and the reference points or interfaces that interconnect the FEs to each other and to functional elements in the public network.

In the above reference model, the LE (local exchange) may represent a different type of edge node function in a fixed network depending on the WLL operator requirements and may include a PSTN switch, a data network router, or a leased line node.

The base station controller provides the connection between the WLL system into the fixed network and controls the base stations and interfaces to the network management unit.

The base station contains the necessary radio receiver and transmitter functions as well as the signaling functions from/to user terminal equipment. It also provides the functions for maintaining the radio path and measuring its performance.

Introduction to WLLs. By Raj Pandya
ISBN 0-471-45132-0

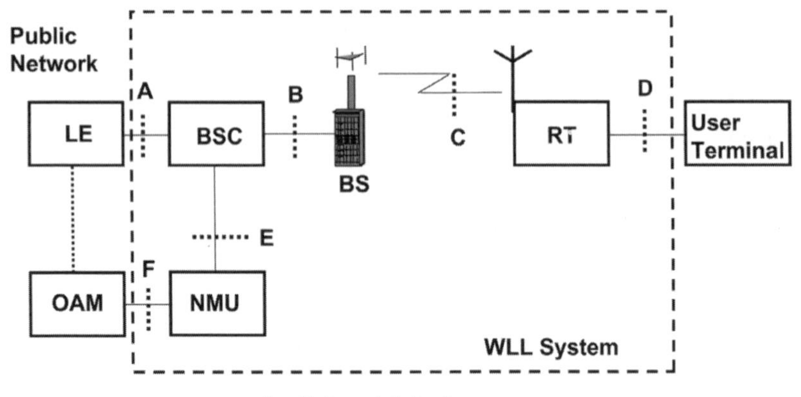

Figure 4.1. General reference model for a WLL system.

The RT (radio termination) provides functions for accessing the radio interface and to support different end-user terminal types (POTS, ISDN, data).

The NMU (network management unit) contains functions necessary for managing data associated with system configuration and customer, system, and radio link parameters. It views the WLL system as a managed entity (ME) within the TMN (telecommunication management network) model.

Interfaces A and C represent the network and radio interfaces, respectively, and are generally the main focus of analysis and decision for planning and deployment of a WLL system. The remaining interfaces (B, D, E, and F) are typically implemented by the equipment manufacturer/provider as proprietary protocols.

Figure 4.2 shows a simplified view of a WLL system which focuses on the major components and interfaces that require more detailed discussion. WLL radio systems at the end-user premises or at the base station consist of many common functions and required physical equipment to support these functions. Thus, there is significant commonality between the radio systems at the end-user premises and at the base station. However, such parameters as transmitter power levels, antenna type and height, and interface requirements (to PSTN) are different for the base station system. Base stations are also required to support additional functions in terms of call control and radio resource management, as well as mobility and call handoff management (if limited or full mobility is to be supported by the WLL system).

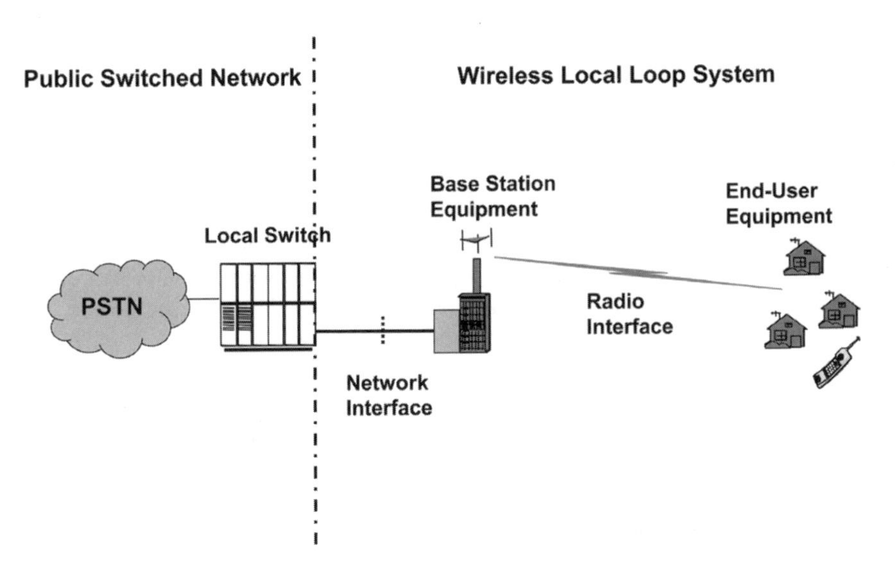

Figure 4.2. Components and interfaces for a WLL system.

4.2 END-USER EQUIPMENT

Figure 4.3 illustrates the basic components of the equipment at one end of a radio link. The three basic components (indoor unit, outdoor unit, and the antenna) are present in all types of radio terminals, independent of whether it is in the local loop (point-to-multipoint) part or in the backhaul radio (point-to-point) part, or whether the terminal is a radio base station or is a radio remote terminal.

In the case of the remote terminal unit, the figure applies to remote units that are not mobile and are installed at the customer premises. If the WLL system supports terminal

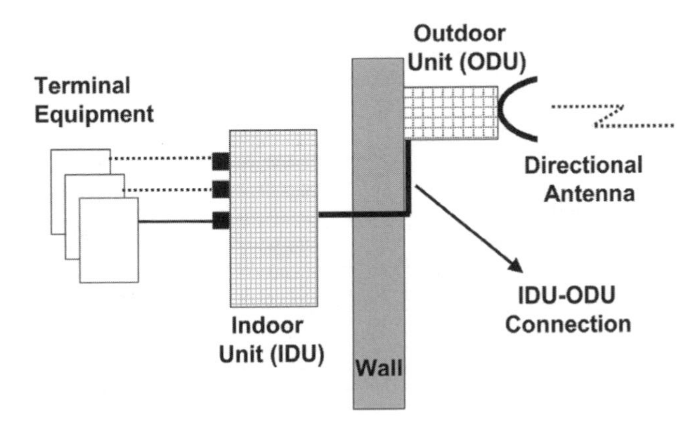

Figure 4.3. Components of a WLL subscriber radio terminal.

mobility, then the remote unit has to be portable, containing all the necessary electronics and power source (battery), and cannot use a directional antenna.

4.2.1 Indoor Unit (IDU)

The basic function of the IDU is to convert the digital signal from the end-users equipment (voice, data, video terminal) into a suitable form for transmission [by the outdoor unit (ODU) and the antenna] over the radio channel and to process the received signals (from the ODU) for presentation to the end-user device. The functions provided by the IDU are shown in Figure 4.4.

The IDU functions comprise the radio transmission and reception functions for the outgoing and incoming user information and control signaling, telemetry, and network management functions for monitoring and configuration of the end-user equipment including the IDU, the ODU, the directional antenna, the power supply for the IDU and ODU, and functions for multiplexing various incoming and outgoing signals (including power supply) over the IDU–ODU feeder cable.

4.2.1.1 Transmitter/Receiver Functions These include functions like source coding, channel coding, interleaving/deinterleaving, and modulation/demodulation as described in Section 2.5. For systems that deploy frequency division multiple access for defining radio channels, generally the modulator in the IDU converts the received signal either to the so-called *base-band* (BB) frequency or to an intermediate frequency (IF). In the case of base-band modulation, the modulating signal corresponds to a frequency that represents the channel bandwidth available on the radio path. This resulting signal, modulated to the available channel bandwidth, is called *base-band* signal.

For example, in the case of GSM, the total available spectrum is divided into 124 FDMA channels, each with a bandwidth of 200 kHz (or 0.2 MHz). The IDU modulator, in this case, will convert the end-user signal into a base-band signal whose center frequency is 100 kHz (or 0.1 MHz). The ODU will then up-shift the

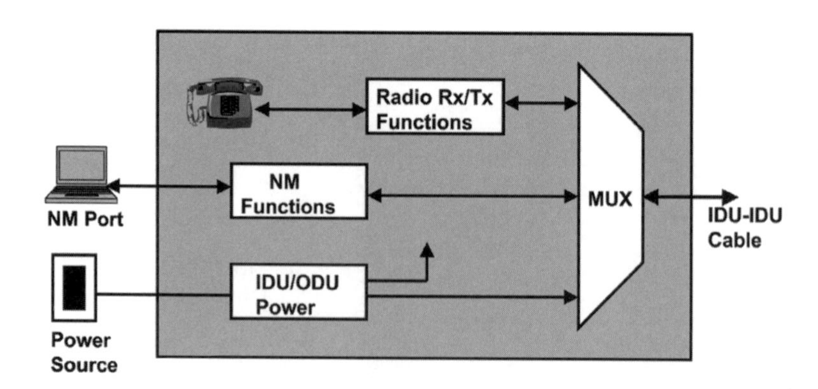

Figure 4.4. Functions associated with the indoor unit (IDU).

base-band signal to the appropriate frequency of the (up-link) channel allocated (by the base station) to the end-user terminal. For example, the 200-kHz up-link channel allocated to the terminal in the GSM example may be in the frequency range 890–890.2 MHz so that the ODU will generate an RF signal for transmission with a center frequency of 890.1 MHz.

Since FDMA channel structure defines a constant channel bandwidth (e.g., 200 kHz in GSM), the same base-band modulating frequency has to be used for both the up-link (IDU → ODU) and the down-link (ODU → IDU) signals. This means that the two signals can not be multiplexed on a single IDU → ODU cable, and two separate cables may be required.

The intermediate frequency is chosen somewhere in between the base-band and the RF center frequency. The IF modulated signal from the IDU is then up-shifted by the ODU to an appropriate RF carrier for transmission. The choice between use of IF versus BB signal between the IDU and the ODU is a design parameter, and the factors that may dictate the choice include

- Better suitability of the IDU modulation method
- Potential for lower loss and interference on the IDU-ODU cable
- Potential for multiplexing all signals between IDU and ODU on a single cable (however, this requires that two different IFs be used for up-link and down-link directions)

4.2.1.2 Network Management Functions

The physical equipment to support network management functions (e.g., terminal control unit and telemetry modem) contains the electronics to monitor the radio terminal as a whole, and it responds to s/w commands from the NM port. It is generally designed to be software-based and to conform with standard s/w and network management interfaces (including SNMP and CMIP) and is capable of being monitored and configured from a remote location. It also provides a communication channel for network management, monitoring, and configuration purposes, which allows the ODU to be controlled from the control terminal in the IDU. At the base station, capability can be provided to diagnose and reconfigure antenna equipment without having to climb the mast.

4.2.1.3 Power Supply

Unlike telephones in a wireline system where a terminal is powered from the local exchange, the electronics involved in a WLL terminal needs to be powered from a power unit at the customer premises. Reliability of the in-house power supply is therefore critical for continuous service. Typically, 48-volt or 24-volt DC power is supplied from the mains supply with a standby battery. Solar power options are also available from some WLL equipment suppliers.

4.2.1.4 Cable Multiplexer

Serves only to share the use of IDU–ODU connection cable between the modulated user data signal, the network management information, and control signals and for provision of power to the ODU.

4.2.2 Outdoor Unit (ODU)

The ODU in a WLL radio terminal basically comprises of the RF transmitter and receiver (transceiver). Its main function is to modulate and amplify the lower-frequency signal received from the IDU into a RF signal at the appropriate frequency for transmission over the radio channel, and to effectively alleviate the effects of signal degradation, and demodulate the received RF signal to a suitable frequency for further processing by the IDU. The ODU functionality does not alter the basic characteristics (like bit rate and synchronization) of the signals received or passed on to the IDU. Figure 4.5 illustrates the functional diagram of a typical ODU of a WLL radio.

4.2.2.1 Receiver Front End The main function of the splitter is to separate the incoming (receive) and outgoing (transmit) signals at the antenna. Before the demodulation stage in the receiver, devices like a low-noise amplifier (LNA) and/or an automatic gain control (AGC) is used to strengthen and stabilize the received signal. The former (LNA) is intended to amplify weak signals while preserving the clarity and quality of the received signal. The latter (AGC) is a variable strength amplifier that is designed to present a constant-strength signal to the receiver. It compensates for unpredictable radio signal losses over the radio channel caused by radio path attenuation that may be caused by adverse weather conditions like rain and snow.

For proper reception of high-bit-rate data or video signals that require high-fidelity reception, an *equalizer* stage may be added before demodulation. The equalizer is intended to compensate for uneven losses or attenuation whereby the

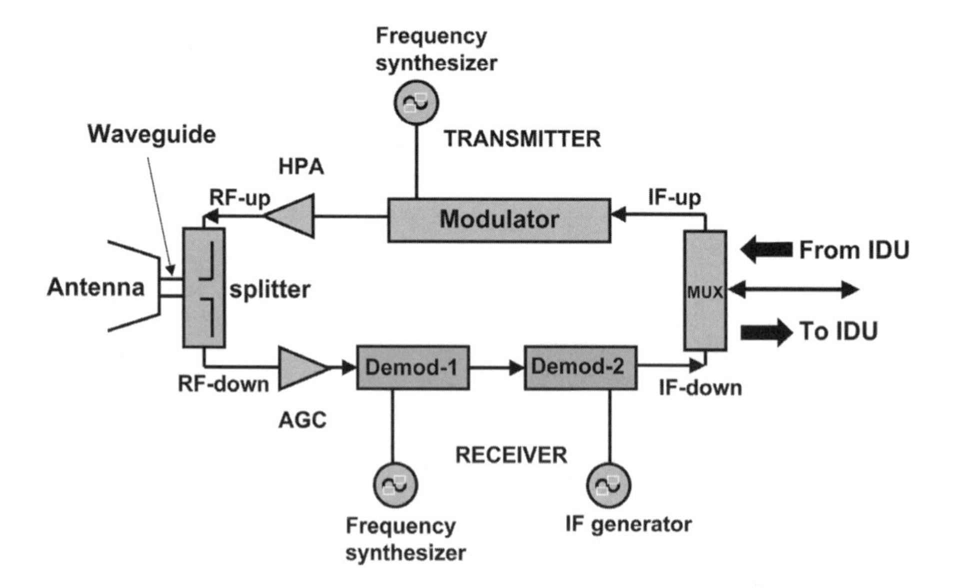

Figure 4.5. Functions associated with the outdoor unit (ODU).

frequencies at the lower end may be affected more (or less) than those at the higher end. These unbalances are more pronounced when the incoming signal is spread over a wide frequency band and is modulated using a higher-order modulation method like 8-QAM or higher.

4.2.2.2 Demodulators The first stage of demodulation in the receiver is used to shift the received signal to the down-link carrier frequency (frequency channel) assigned to the end-user terminal (by the base station). This shifting is achieved by using a frequency synthesizer that allows the receiver to be tuned to any of the down-link frequency channels defined for the operating frequency band allocated to the system (e.g., down-link channel 1 in GSM with the center frequency = 935.1 MHz). The second demodulation stage then down-shifts the signal to the appropriate intermediate frequency (IF-down) which is then passed on to the IDU for further processing. Thus the frequency synthesizer is a highly reliable tunable oscillator that is software-controlled, and the IF generator is a very stable and accurate fixed frequency oscillator.

4.2.2.3 Modulator The transmit part of the ODU works like a receiver in reverse. The up-link signal from the IDU is received over the IDU–ODU cable connection. It is the user signal modulated by the up-link intermediate frequency (IF-up) which is then up-shifted to the appropriate up-link RF channel using the frequency synthesizer. The frequency synthesizer ensures that the outgoing RF signal is modulated by the up-link frequency channel assigned by the base station. The resultant signal is then amplified at the high-power amplifier (HPA) and then conveyed by means of the splitter and waveguide to the antenna for transmission.

4.2.3 Performance Parameters for the Outdoor Unit

The quality of communication (perceived by the end user), operational suitability, and reliability of the end-user terminal depends on a number of parameters associated with the ODU. Some of these parameters include

- Transmit power
- Receiver sensitivity
- Frequency stability
- Adjacent channel and co-channel interference
- Spurious emissions
- Transceiver failure rate (MTBF)

4.2.3.1 Transmit Power This may be specified in mW or dBm (decibels relative to 1 mW) with typical values ranging between 10 and 30 dBm. The transmit power in WLL systems is generally set at a level to ensure good range during a specified level of signal loss on the radio channel, but at the same time to minimize the possibility of causing unwanted interference with adjacent radio systems when there is little signal loss over the radio channel. In cellular systems, a major consideration in setting the terminal output power is battery conservation, and automatic transmit power control (ATPC) methods are utilized to continuously monitor and adjust the transmit power. If warranted, similar ATPC methods can also be deployed in WLL systems.

4.2.3.2 Receiver Sensitivity The range of the radio system is also limited by the minimum receiver sensitivity. The receiver sensitivity threshold is measured in terms of received signal level (RSL) measured in dBm that will ensure a specified maximum bit error rate (BER). For example, the maximum BER level used is generally 10^{-6}, and typical levels for receiver sensitivity in WLL systems range between -70 dBm and -95 dBm. Receiver sensitivity is affected by the type of modulation scheme used. High-bit-rate modulation methods like 4-, 8-, 16-, and 64-QAM may conserve spectrum but have a deleterious effect on required receiver sensitivity. A maximum limit is generally set for the received signal level in order to ensure proper operation of the receiver and to protect the receiver from damage due to overload.

4.2.3.3 Frequency Stability This measures the stability and accuracy of the oscillators used in the modulators/demodulators so that the actual center frequency of the selected radio channel does not deviate more than $X\%$ (where the value of X is typically 0.0015).

4.2.3.4 Adjacent Channel and Co-channel Interference Parameters These measure the ability of the receiver to tolerate interference from signal power from an adjacent frequency channel and from a channel using the same frequency in a distant cell, respectively, without affecting the receiver's ability to maintain the target BER. These are generally specified as carrier-to-interference (C/I) ratios for a given BER target. For adjacent channel interference to be a problem, the signal level from the adjacent channel has to be sufficiently strong. On the other hand, even a very weak co-channel signal can cause unacceptable interference. The following specifications are typical:

$$\text{C/I (adjacent channel)} = -10\,\text{dB} \qquad \text{for BER} = 10^{-3}$$

$$\text{C/I (co-channel)} = +13\,\text{dB} \qquad \text{for BER} = 10^{-6}$$

4.2.3.5 Spurious Emissions Parameter This measures unwanted signals transmitted by the radio transmitter outside of the radio band. Standards for spurious emissions are set by appropriate regional or national regulatory bodies, and they generally maintain a strict regime of specification and conformance. The unwanted emissions are strongest at the harmonic frequencies and should be monitored at least up to the third harmonic of the operating frequency band.

4.2.3.6 Transceiver Failure Rate (MTBF) This is primarily determined by the reliability of the ODU, which is generally specified by the mean time between failures (MTBFs). The target for a radio link failure is on the order of MTBF = 10 years, so that the MTBF for the individual systems at the two ends of the radio link needs to be 20 years. Since MTBF is not an easily measurable parameter, it is generally based on theoretical calculations that take into account the reliability of the individual system components.

4.3 THE BASE STATION SYSTEM

The other major component in a WLL system is the base station system (BSS) whose chief function is to provide radio coverage for end users within a single or multiple cells. Radio traffic passes from the end-user terminal to the BSS in the up-link direction and from the BSS to the terminal in the down-link direction.

The BSS manages all aspects of the RF links. It allocates the radio channels (frequency and time slot in TDMA systems and frequency and spreading codes in CDMA systems) for each terminal, and it controls the power level the terminal should use. During the call, it monitors the radio channel and releases it when the call is completed. Other ancillary functions of the BSS relate to efficient functioning of the cells. These functions include the definition of radio channels into traffic and signaling channels and collection and analysis of signal quality measurements. Thus the key functions performed by the BSS include

- Radio channel management
  - Configuration of radio channels
  - Channel selection, allocation, release
  - Channel blocking indication
  - Monitoring of idle channels
- Automatic transmit power control (if implemented)
- Digital signal processing for
  - Transcoding and rate adaption unit (TRAU)
  - Channel coding and decoding
  - Termination of control channels

In typical WLL implementations, the base station system is divided into two functional blocks: base station controllers (BSCs) and base transceiver station (BTSs) or base station (BS). In a hierarchical architecture shown in Figure 4.6, the BSC generally serves a number of BTSs which are located at the cell sites where each BTS provides radio coverage for a number of end-user terminals within the cell.

The BSS functions in the architecture shown in Figure 4.6 are therefore partitioned into the BSC and the BTS. The allocation of BSS components and functions between the BDC and the BTS are summarized below.

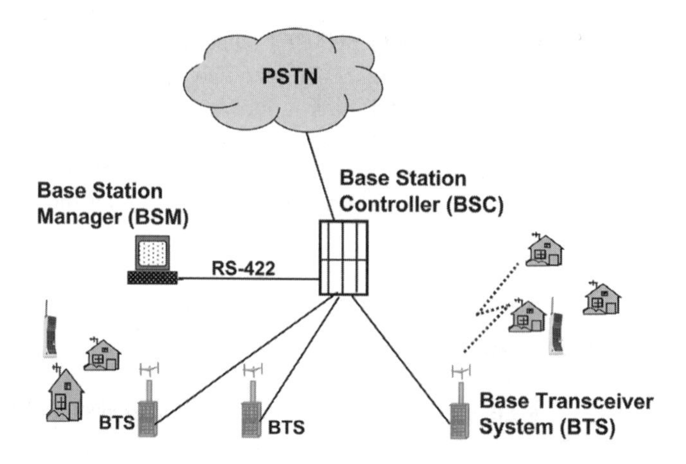

Figure 4.6. Typical base station system (BSS) architecture.

BSC Components and Functions	BTS Components and Functions
BSC control processor	Radio transceivers (modulation/demodulation)
Signaling links to PSTN	Equalizers
Control of base transceiver stations	Channel coding units
Radio resource management	Interleaving/deinterleaving
Handover support (for mobility)	Transcoding and rate adaption (TRAU)
Call processing between WLL network and PSTN	BTS control processor
	Signaling links to BSC
	Combiners
	Antennas

Though the above list provides a broad allocation of functions to the BSC and BTS, the actual allocation could vary depending on the individual implementations. For example, the BSC in some implementations may support switching and service support functions—especially when the WLL operator wishes to minimize interconnection charges to the PSTN operator for call completions within the WLL network. Since the BSC and BTS equipment is generally supplied by the same vendor, the BSC-to-BTS interface is a proprietary interface designed for maximum efficiency.

Whereas Figure 4.6 represents a hierarchical architecture, some WLL network implementations utilize a distributed architecture where each individual cell is served by its own base station which also provides direct links to the PSTN local exchange. The choice between a hierarchical versus distributed network architecture is generally dictated by the size and geographic distribution of the subscriber base and the associated economics.

4.4 RADIO INTERFACES FOR WLL SYSTEMS

4.4.1 Radio Interfaces Based on Current Cellular/Cordless Standards

The radio channel that provides a communication path between the end-user terminals and the base station is governed by the radio interface deployed for the WLL system. The radio interface specifications not only define the frequency band of operation and the duplexing (FDD/TDD) and multiple access (FDMA/TDMA/CDMA) methods, but also specify the detailed channel rasters, framing structures, timing requirements, source and channel coding schemes, and modulation methods.

The radio interfaces used in WLL systems fall in two basic categories: those based on current and emerging cellular and cordless telecommunication system standards or those which are proprietary and are developed by individual WLL equipment vendors to meet specific applications and markets.

The advantage of WLL systems based on cellular or cordless radio standards is that these systems are manufactured by cellular or cordless system vendors as a byproduct of their main business, with resulting potential for lower cost and easy availability of WLL equipment. However, these systems need to operate in the frequency bands standardized for the cellular or cordless applications, and the frequency band may already be in use in a geographic area where the WLL system is planned to be deployed. Alternatively, if the frequency band is available, the licensing fees for the band may be too high for the WLL business case to be viable.

It is possible to utilize radio interfaces based on first-generation analog cellular systems operating in the 450-MHz band (e.g., NMT-450) for WLL systems. Since these analog cellular systems are now almost all replaced by second-generation digital cellular systems like GSM, the frequency band may be available for use in WLL applications. Because of the lower frequency of operation, WLL systems based on NMT-450 also have the advantage of larger cell size and wider coverage due to lower path losses. However, deployment of these systems poses the potential risk of the WLL operator being locked into an obsolete technology for which equipment for replacement or expansion purposes may not be easily available.

Major digital cellular and cordless systems like GSM, IS-136 (DAMPS), IS-95 (CDMA), DECT, and PHS, which are commonly used as the basis for WLL systems, were described in Chapter 3. Table 4.1 provides a summary of these standards. These cellular and cordless radio standards represent the so-called second-generation (2G) mobile and personal communication systems developed and deployed since the beginning of the 1990s. In the meantime, work has been underway in the regional and international standards organizations to specify new and evolved radio interface standards to meet the needs of the new or third-generation (3G) mobile communications. These radio interfaces for 3G systems were briefly discussed in Section 3.9.

Third-generation mobile communication systems are already being planned around the world, and some initial systems based on W-CDMA and CDMA 2000 are now commercially available. Commercial WLL systems based on these new or

TABLE 4.1. Summary of Digital Cellular and Cordless Standards Used in WLL Systems

Radio Standard	Frequency Band (MHz)	Access Method	Reference Standard	Relevant SDO[a]
GSM (900)	890–915 935–960	TDMA/FDD	ETS-300, 500, 700 series	ETSI (Europe)
GSM (1800)	1710–1785 1805–1880	TDMA/FDD	ETS-300, 500, 700 series	ETSI (Europe)
GSM (1900)	1850–1910 1930–1990	TDMA/FDD	ANSI-007A	Joint T1/TIA (U.S.A.)
DAMPS (800)	824–849 869–894	TDMA/FDD	IS-136	TIA (U.S.A.)
DAMPS (1900)	1850–1910 1930–1990	TDMA/FDD	ANSI-009, 010, 011	Joint T1/TIA (U.S.A.)
CDMA (800)	824–849 869–894	CDMA/FDD	IS-95	TIA (U.S.A.)
CDMA (1900)	1850–1910 1930–1990	CDMA/FDD	ANSI-008, 018,019	Joint T1/TIA (U.S.A.)
DECT	1880–1900	TDMA/TDD	ETS 300 175	ETSI (Europe)
PHS	1895–1918	TDMA/TDD	RCR-28	RCR (Japan)

[a]Standards Development Organization

enhanced CDMA technologies are being developed by many vendors and are expected to be available for deployment in the very near future. In fact, some vendors are already offering WLL systems based on these technologies.

4.4.2 Radio Interfaces Based on Proprietary Radio Technologies

There are a significant number of proprietary WLL systems which utilize radio interfaces that do not conform to any existing radio standards and are designed to operate in frequency bands that are not currently assigned to cellular or cordless applications. As mentioned earlier, the advantages of such proprietary systems is that they are designed specifically for WLL applications (rather than being a byproduct of existing cellular/cordless systems), so that they can provide better efficiency and coverage and can accommodate frequency bands that are different from those already in use for existing cellular or cordless systems. Examples of these proprietary interfaces and their radio characteristics are summarized in Table 4.2. These proprietary systems are described in greater detail in Chapter 7. A W-CDMA WLL radio interface is likely to become the South Korean national standard under the auspices of TTA (Telecommunications Technology Association), the South Korean telecommunications standards organization.

TABLE 4.2. Radio Interface Characteristics for Some Proprietary WLL Systems

Parameter	Air Loop System (Lucent)	Internet FWA (Nortel)	SWING (Lucent)	W-CDMA WLL (LGE)
Access method	DS-CDMA	TDMA	TDMA	DS-CDMA
Duplex method	FDD	FDD	TDD	FDD
Frequency bands options	3.4–3.45 GHz 3.5–3.55 GHz, or 3.45–3.5 GHz 3.55–3.6 GHz	3.4–3.6 GHz	1.88–1.9 GHz 1.9–1.92 GHz 1.91–1.93 GHz	2.3–2.33 GHz 2.37–2.4 GHz
Duplex separation	100 MHz	100 MHz	N.A. (TDD)	70 MHz
CH bandwidth	5 MHz/FA	2 MHz	2 MHz	10 MHz/FA
Total RF channels	115 (max)	264 (max)	120	500 (max)
Modulation	QPSK	pi/4-DQPSK	GFSK	QPSK
Speech coding options	LD-CELP-16 kb/s ADPCM-32 kb/s PCM-64kb/s	ADPCM	ADPCM PCM	ADPCM

4.4.3 Protocols at the Radio Interface

An interface represents the point of contact between two adjacent entities—for example, the end-user terminal and the BTS in the case of a radio interface. However, the radio interface can carry information not only between the radio terminal and the BTS, but also between the radio terminal and the BSC and the MSC (or local switch) to support the required functions. These messages from the radio terminal (to BSC or MSC/LE) are carried transparently through the BTS. In cellular mobile systems, there may also be message flows between the radio terminal and the HLR for support of supplementary services. The flow of information across the radio interface in a cellular mobile system or a WLL system supporting mobility is shown in Figure 4.7.

The layered architecture for the radio interface and the functions supported at the different layers are illustrated in Figure 4.8. Within the layered protocol architecture, the radio channel with its key characteristics described in the preceding few sections represents layer 1 or the physical layer. Layer 2 is required to support the acquisition and control of the radio link and is generally partitioned into media access control (MAC) and link layer control (LLC) functions. The LLC is generally based on a modified form of LAPD (link access protocol for D channel in ISDN) to support radio interface specific features. As illustrated in Figure 4.8, Layer 3 is divided into three sublayers that deal with radio resource (RR) management, mobility management (MM), and call management (CM), respectively. Radio resource

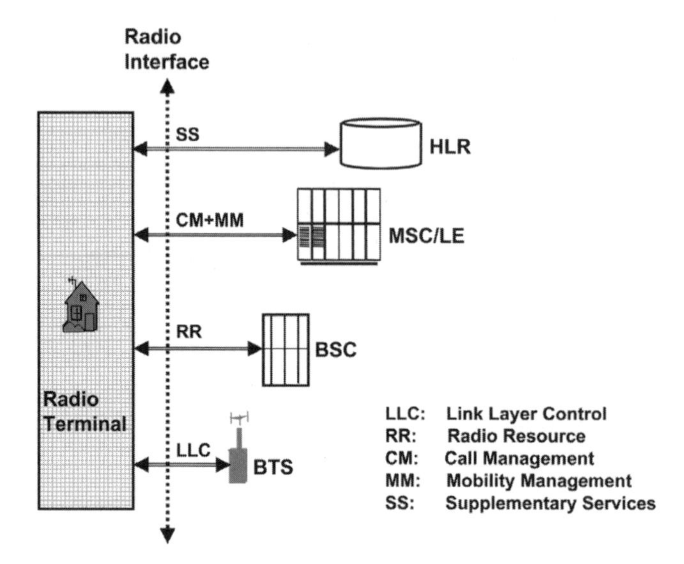

Figure 4.7. Functions supported by the radio interface protocols.

management is concerned with managing logical channels—their assignment and performance measurements. Mobility management functions like terminal registration, terminal location updating, and authentication are required where terminal mobility is to be supported. Call management sublayer functions are concerned with call and connection control, establishing and clearing call/connections, and management of supplementary services.

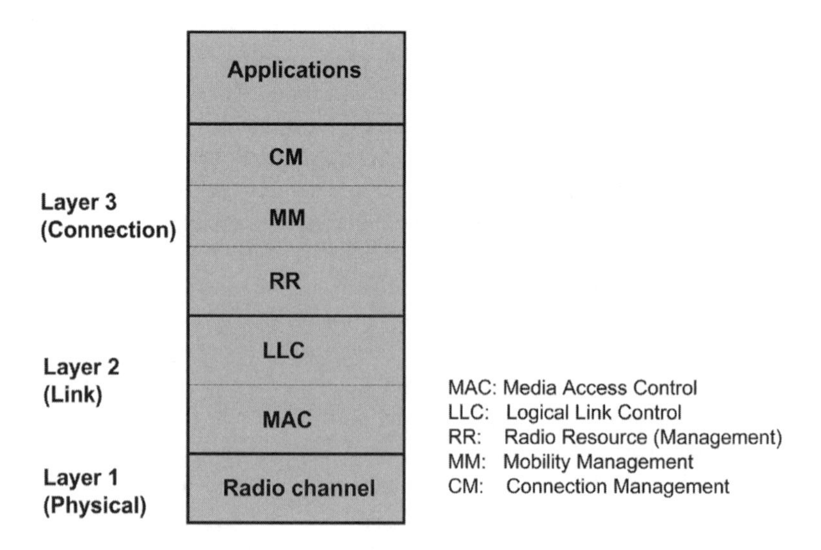

Figure 4.8. Layered architecture for the radio interface.

4.5 NETWORK INTERFACES FOR WLL SYSTEMS

4.5.1 Transport Options

As illustrated in Figure 4.1, the WLL network interface specifies the transport and protocols associated with the base station to PSTN local switch connection, which is generally referred as the *backhaul* component. The various transport and protocol options available for the design of this link for WLL systems are shown in Figure 4.9.

Coax cable: This may be a viable and perhaps economic option (as leased lines) in metropolitan and urban applications. However, unless good-quality coax cable connectivity is available, it may be difficult to meet service quality targets.

Fiber links: Again this may be viable only in metropolitan and urban environments (where fiber connectivity is available). The transmission quality and reliability of this transport medium will be excellent.

Microwave Links: These are frequently deployed in a point-to-point mode at a range of frequencies. The coverage range of microwave systems reduces as the frequency of operation becomes higher, implying that more microwave links may be required to cover a given distance. However, higher-frequency microwave systems have some cost advantages in terms of smaller dishes and smaller antenna structures. Since these are radio links, their availability and reliability are critical design factors.

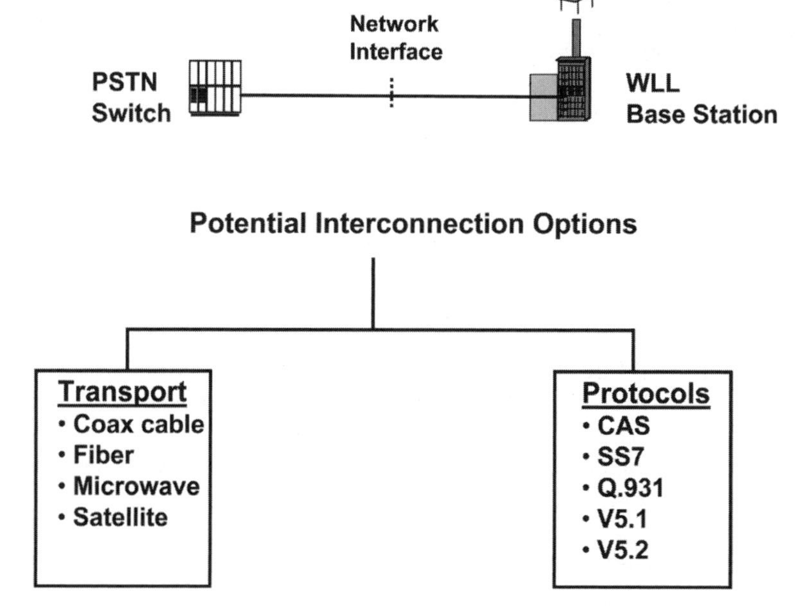

Figure 4.9. Interconnection of WLL to PSTN: transport and protocol options.

Satellite Links: Though leasing satellite links is rather expensive, VSAT (very small aperture terminals) may be economically viable to connect very remote base stations to the PSTN. However, the large propagation delay involved is generally an undesirable feature.

4.5.2 Protocol Options

Independent of the transport medium used to connect the base stations to the PSTN switch, a common protocol needs to be supported at both ends of the link. If the base station and the switch are provided by the same vendor as a package, the vendor may implement a proprietary interconnection protocol. Such proprietary protocols can be more efficient and can save bandwidth over the link. However, the down side is that the operator is locked into the products provided by the vendor and will have little flexibility in using other vendors' products (even if they are more economical).

In WLL systems where the base stations are to be connected to existing PSTN switches, there are a number of standard protocols that can be deployed. The choice will generally depend on the additional cost for upgrading the PSTN switch (if necessary) and the range of services that the WLL system needs to support.

CAS (channel-associated signaling) can only carry simple voice traffic and thus limits the range of services available to the WLL subscribers. It further requires dedicated circuits to be assigned to each subscriber and is therefore very inefficient in terms of resource utilization.

Q.931 or DSS1 is the ISDN user–network interface (UNI) protocol, and some WLL systems based on cordless radio (e.g., PHS) utilize Q.931-based interface to the PSTN. It can only provide services supported by the basic rate ISDN (2B + D) link.

SS7 (signaling system 7) is the network-to-network interface (NNI) protocol deployed in most modern digital networks. Many cellular systems utilize SS7-based interfaces between the BSC and MSC (generally known as the A interface). Many WLL systems based on cellular standards (e.g., GSM) will support SS7 for backhaul connection to the PSTN.

V5 (V5.1 and V5.2) interface is an open interface specially developed and standardized (by ETSI) between the local exchange (LE) and an access network (AN). The access network may be a remote switching unit (RSU), a private branch exchange (PBX), or a radio access network (RAN). The BSC in a WLL application represents a radio access network, and a V5 interface can be used to advantage in WLL system. The V5.1 interface requires one time slot per subscriber and offers no concentration or *trunking* of time slots. The V5.2 interface, on the other hand, allows time slots to be *trunked* or shared among subscribers on a demand basis which can lead to significant cost savings. The ITU version of the V5.2 interface is described in ITU-T Recommendation G.965; and the North American version, known as the TR303.V5.2 interface, is described in greater detail in the next section.

4.5.3 V5.2 Interface for Interconnection of WLL Systems

The V5.2 interface is designed to deliver telecommunication services supported by the local exchange to WLL subscribers served by a BSC irrespective of the supplier of the BSC and related equipment. Not only does the V5.2 interface frees up the WLL service provider from being locked into a proprietary switch interface for network interconnection, but the service provider is ensured of the local exchange services to be delivered in a flexible, efficient, and transparent manner. A list of functions that are typically supported by the local exchange across the V5.2 interface includes

- Call processing.
- Call records and billing
- Numbering plan management
- Supplementary services
- Answer supervision
- Time slot allocation and management

The radio access network or the BSC is only responsible for such functions as call delivery to and call initiation from subscribers, radio resource allocation and managementary, management of end-user terminals, subscriber paging, and subscriber authentication.

Most modern digital local switches are equipped to support the V5.2 interface. However, if the local switch is not equipped to support the V5.2 interface, a so-called network interface unit (NIU) may need to be inserted between the WLL base station and the PSTN switch. The NIU then communicates with the BS using V5.2 protocol, and it communicates with the PSTN switch using the type of signaling supported by it (e.g., SS7 or R2 signaling). Some of the advantages of using a V5.2 interface in WLL applications are as follows:

Equivalent Service Level: V5.2 allows transparent delivery of PSTN or LE functions and services such as billing, numbering plans, and supplemental services by the switch (such as call forwarding, call waiting, three-way calling, etc.).

Flexibility for Line Expansion: V5.2 can support different types of access arrangements including PBXs and remote switching units, so that the service provider has the option of expanding service using multiple access arrangements.

Choice of Multiple Vendors: V5.2 open interface provides many choices in vendors (and technologies) for the WLL system, thereby allowing an increased choice in the service portfolio potential for competition in prices.

Trunking Efficiency: Since V5.2 is a trunked interface, it allows time slots to be allocated and shared between subscribers on a per-call basis. This approach is more cost-effective in terms of interconnection links and is more reliable and robust in terms of link failures.

Minimizes RAN Functionality: V5.2 allows clear partitioning of functions between the LE and the RAN, whereby the RAN needs to support only those

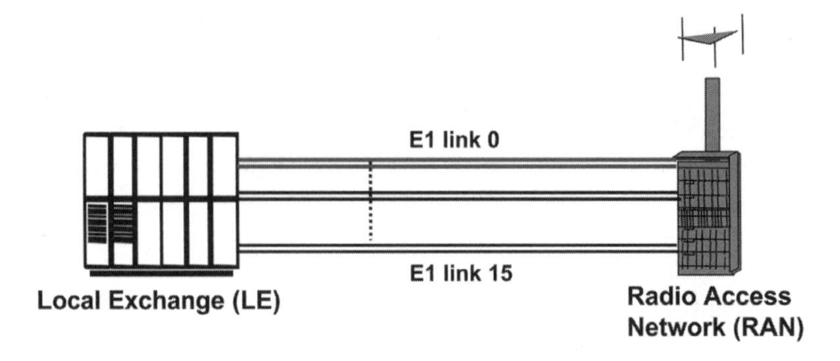

Figure 4.10. Route architecture for the V5.2 digital interface.

call delivery functions required by radio access. This leads to simpler design for the RAN equipment and associated cost reductions.

Figure 4.10 illustrates the V5.2 connectivity with active standby common control signaling utilizing the V5 protocol. Each V5 route consists of 16 E1 links (E1 link = 2.048 Mb/s) that support 32 time slots each. Time slot 0, as per E1 specification, is used for synchronization and alarms. The first two E1 links contain one common control link, one is active and the other is on standby, each consisting

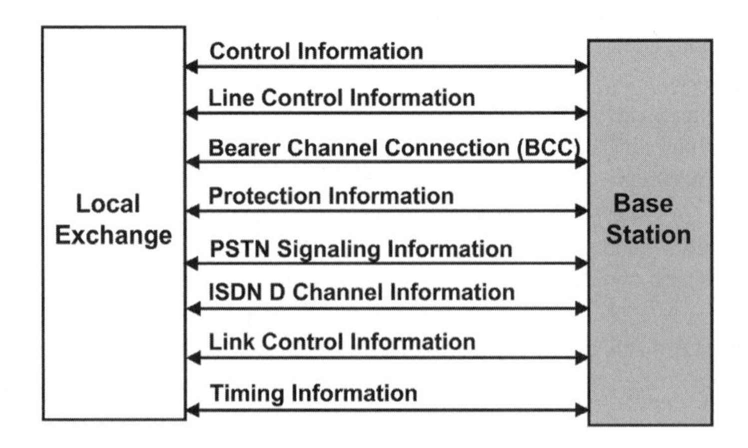

Figure 4.11. Protocol functions for the V5.2 digital interface.

of one E1 time slot (time slot 16). The remaining time slots and E1 links are used for dynamically allocated voice channels. A total of 494 voice trunks are available (16 E1s × 31 time slots less 2 control channels). Where V5.1 multiplexes a single time slot to each user, V5.2 adds concentration, which dynamically allocates a time slot to each user.

Thus, in V5.2 operation, 16 E1 connections are pooled together to form a single group providing a V5 route with 494 potential connections. Base station equipment is available which can support up to four V5 routes, for a total of 64 E1 spans.

The V5.2 interface provides the system with greater operational control and functional efficiency. V5.2 supports more subscribers per E1 link and thus can be implemented at a lower total cost and is thus more efficient than V5.1 and many other types of non-V5, nonconcentrating PSTN interfaces. V5.2 also provides for increased fault tolerance which results in overall improved quality and reliability for the customer.

The protocol functions for the V5.2 interface are shown in Figure 4.11. The bearer channel control (BCC) protocol is used to assign bearer channels under the control of the local exchange (LE). A link control protocol is defined which supports the management functions of the 2.048-Mb/s links of the V5.2 interface. The protection protocol supports switching of logical control channels. Control protocol manages information related to the operational state of supported user ports (e.g., PSTN, ISDN, leased circuits, etc.). PSTN and ISDN protocols transfer information about the analog and ISDN line states, respectively, over the V5.2 interface to the LE that supervises the call.

CHAPTER 5

Radio Access Characteristics and Radio Planning

The quality of service perceived by the end user, as well as the reliability of the link in a radio system, is highly dependent on the conditions associated with the radio path. Depending on the frequency of operation, the radio path may be subject to various types of radio interference and signal fading. In order to maximize the coverage and capacity of WLL systems, it is necessary to pay special attention to the planning and design of the radio access network. Frequency planning, placement of base station and subscriber antennas, and design and configuration of base station antenna systems are some of the factors that will significantly impact the efficient and effective operation of a WLL system. This chapter addresses these radio-related aspects of WLL system design and planning.

5.1 INTRODUCTION TO RADIO DESIGN FOR WLL SYSTEMS

The outside plant in a wire-line access or distribution network consists of cable and twisted copper wire that connect individual end-users' equipment to the local exchange in order to deliver telecommunication services. The reliability of the connection and the quality of service provided to the end user is essentially assured within very narrow tolerances. In the case of the radio access network for WLL systems, the reliability and quality of the connection between the end-user equipment and the local exchange is dependent on (a) the prevailing radio propagation conditions along the radio path and (b) a number of radio-related parameters associated with radio access and end-user equipment. The planning and design of the radio access network for WLL systems is very much different and more complex than that for wire-line distribution networks. Such aspects as radio coverage, radio range, frequency reuse, and radio path loss need to be addressed, which further require close attention to (a) the type and orientation of base station and end-user antennas and (b) the optimum location and power outputs of base station transmitters and the level of receiver sensitivity at the end-user sites. Radio

Introduction to WLLs. By Raj Pandya
ISBN 0-471-45132-0

system planning and design of a WLL system requires some understanding of issues related to

- Modes of radio wave propagation for the deployed frequency band
- Potential sources of signal loss over the radio path
- Frequency planning and frequency reuse
- Factors affecting WLL system capacity and range
- Capacity and coverage expansion techniques

5.2 MODES OF RADIO WAVE PROPAGATION

Radio waves are a form of electromagnetic radiation, and they may be reflected, refracted (slightly bent), diffracted (slightly swayed around obstacles), and scattered. Radio waves can therefore take one or more paths to travel between a transmitter and the receiver as a result of these phenomena. Table 5.1 summarizes the radio propagation phenomenon, the resulting radio paths, and examples where they play a key role. It needs to be emphasized that depending on the frequencies being used in a WLL system, the received signal may travel over multiple paths using different propagation phenomena.

The propagation characteristics of radio waves are highly dependent on the frequency band in use as well as on the topography and terrain over which the radio waves need to travel. Depending on the frequency of operation, the propagation characteristics may also be affected by the prevailing climatic conditions like rain and snow.

Many of the WLL systems use frequencies close to or above 1 GHz and are therefore generally restricted to line-of-sight (LOS) operation. LOS systems are generally restricted to short-haul applications because of the constraint imposed by the curvature of earth's surface. Depending on the associated terrain and antenna

TABLE 5.1. Modes of Radio Wave Propagations and Associated Applications

Radio Wave Path	Propagation Type	Typical Applications
Line-of sight (LOS) path	Direct	WLL and microwave systems that typically use frequencies above 1 GHz
Surface wave path	Diffraction or bending along the earth's surface	Medium-wave and long-wave systems used for AM/FM broadcasting and some very low frequency military systems
Troposphere scatter path	Reflection from the troposphere	Applications in the UHF band
Sky wave path	Refraction/deflection from the ionosphere	Short-wave radio applications in the VHF band

heights, the maximum range may vary from 15 to 50 km. In order to maximize received signal levels in a WLL system, operating in an LOS mode, it is necessary that the antenna at the end-user sight is in the LOS of the antenna at the radio base station serving the end user.

Though line-of-sight is the primary path of signal transmission for systems operating above 1 GHz, the other phenomena such as surface wave diffraction, tropospheric scatter, and ionosheric refraction also contribute toward the received signal. However, these latter contributions to the received signal can lead to overall signal degradation that will vary with the operational frequency band.

5.3 POTENTIAL SIGNAL LOSSES IN THE RADIO PATH

There are many factors that can affect the overall received signal level and quality over the radio path. For WLL systems that typically deploy frequencies above 1 GHz, the major contributing factors to signal loss may include

- Fading (absorption) due to precipitation
- Attenuation due to ground coverage (buildings, vegetation)
- Signal absorption due to atmospheric gases
- Multipath fading due to reflection from obstacles
- Multipath fading due to refraction from within the earth's atmosphere

Which set of factors dominate the overall signal loss depends on the actual frequency band that is being used for the radio access system.

In order to estimate the signal loss over an LOS radio path in a WLL system, one starts from the so-called *free-space path* which represents a direct path during ideal weather conditions when the effects of atmospheric disturbances and climatic conditions are minimum. Given the frequency of operation, typical range (distance between transmit and receive antennas), nominal transmit antenna power, and antenna gains at the two ends, one can estimate the *received signal level* (RSL) in dBm for free-space path. ITU-R Recommendation P.525 provides guidance for estimating free space signal loss on a radio path.

The difference between the free space RSL and the nominal receiver sensitivity provides the loss margin that can be tolerated due to various signal loss factors listed above during typical operating conditions. ITU-R Recommendation P.530 may be used as a reference for guidance and formulas to estimate path losses due to such climatic factors as rain fading, snow, fog, and so on; signal absorption due to atmospheric gases; attenuation due to terrain conditions; and multipath fading.

5.4 MULTIPATH FADING AND FRESNEL ZONES

In WLL systems, which typically depend on LOS operation, the primary source of signal loss or degradation is multipath fading. Multipath fading is caused when the original signal combines at the receiver with a reflected signal from an obstacle in a way to cancel part of the original LOS signal. The phenomenon of multipath fading is illustrated in Figure 5.1.

As shown in the Figure 5.1, if an obstruction causes a reflection that produces an alternate reflected path which has a length equivalent to the direct path plus a distance equal to odd multiples of half the radio signal wavelength, then the two signals will arrive at the destination in antiphase. This will produce interference causing significant attenuation or fading of the original direct path signal with reduced net signal level at the receiver. This type of fading is referred as *multipath fading*.

The simplest way to avoid multipath fading caused by reflections from obstacles in WLL systems is to be thorough during the link planning phase of the service provisioning process and ensure that LOS path between the base station and subscriber unit antennas is clear and that there are no potential obstacles inside the first *Fresnel zone*. Figure 5.2 illustrates the concept of Fresnel zones.

As illustrated in Figure 5.2, the first Fresnel zone (generally referred simply as the Fresnel zone) is an elliptically shaped three-dimensional volume surrounding the direct LOS radio path. The (first) Fresnel zone corresponds to the volume within the ellipsoid with a path length difference of one-half wavelength so that reflection near the outer boundary of the Fresnel zone will cause serious interference with direct signal transmissions and result in signal loss. The diameter of the Fresnel zone is dependent on the length of the link and on the frequency of the carrier to be transmitted over the link.

Formulas to calculate the diameter of the Fresnel zone are available so that the link planner can determine whether the Fresnel zone is free of obstacles (e.g., see *Wireless Access Networks* by Martin C. Clark, John Wiley & Sons). The Fresnel zone should be kept clear of any obstacles so that destructive radio reflections from objects within the zone do not lead to serious multipath fading and resulting signal loss.

Note that multipath fading can also be caused by alternate transmission paths of different lengths caused by refractions within the earth's atmosphere as well as by

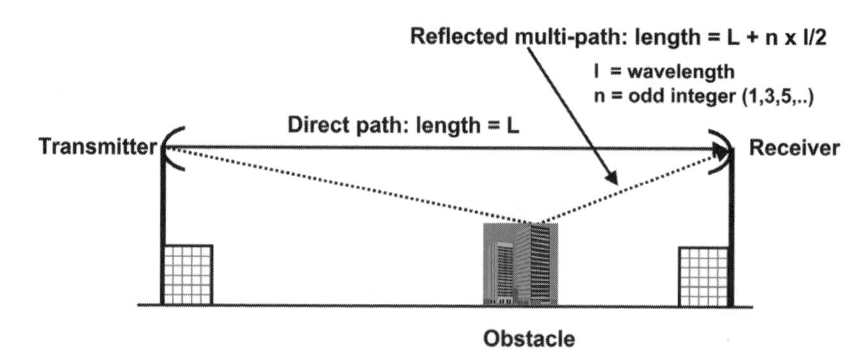

Figure 5.1. Multipath signal loss due to reflection from and obstacle.

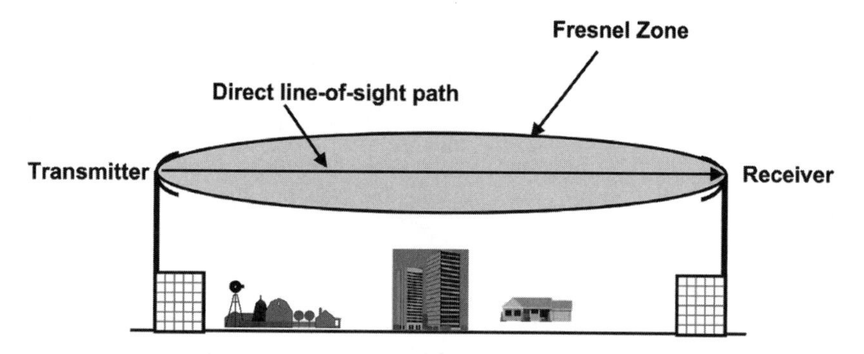

Figure 5.2. Concept of Fresnel zones in radio communications.

reflections from a large body of water. Generally, the best way to avoid multipath fading effects is to choose a transmit antenna site and height which ensures a clear LOS path to the potential receiver locations with no obstacles in the Fresnel zone to cause reflections.

5.5 AVAILABILITY AND FADE (OR LOSS) MARGIN FOR A RADIO LINK

In a wired telecommunication system the end user is guaranteed the availability of a connection and its quality for the duration of the communication. However, in a radio system there are a number of tangible and intangible factors that affect the received signal. The tangible factors include the power output of the transmitter and the sensitivity of the receiver. The intangible factors are the effects of the prevailing radio propagation conditions on the received signal. As discussed in the preceding section, these propagation effects may include signal absorption, signal attenuation, and signal fading.

Under these conditions it is not easy to specify the *reliability* of the radio communication link provided to the end user. The only way to address this problem is to focus on the quality of service *available* to the end user—generally in terms of the bit error rate (BER) of the digital signal observed at the receiver. The BER observed across a radio link varies with time because of the changing propagation conditions. In order to assess the reliability of a radio link, it is therefore necessary not only to set a threshold for the BER, but also to look at how often and for how long this threshold is violated.

Thus, the *availability* of a radio link is generally specified as the percentage of time the BER remains above a given threshold value. For example, the *availability* of a radio link may be specified as 99.99% for a target BER of 10^{-6}.

The *range* of a radio system is measured in kilometers and is intended to provide an estimate of the coverage provided by the radio system. The *range* of a radio system is specified for maximum allowable transmit power, minimum threshold received signal level (RSL), and a given target availability (for specified BER).

The radio system is designed so that the received signal level is well above the threshold RSL and the system operates with few errors. The *link budget* is defined as the maximum allowable signal loss on top of the *free-space loss* for the radio link. The difference between the *free-space level* of attenuation and the level of attenuation that will bring the received signal strength to the threshold receiver sensitivity is called the *fade margin*, and the fade margin represents the extra allowable level of signal loss that can be tolerated before the system is considered unavailable.

Radio link planning is generally based on a given minimum fade margin that is set by adjusting the transmitter power output. Large fade margins are, however, undesirable because under normal operation they can result in interference with neighboring systems operating in the same frequency bands. Sometimes automatic transmit power control (ATPC) is employed to automatically control transmitter power output based on the prevailing fade or propagation loss conditions.

5.6 PLANNING FOR LOS PATHS IN WLL SYSTEMS

As discussed in preceding sections, the signal loss on the radio link is caused by intangible causes (atmospheric, climatic, and geographic) as well as tangible causes (obstacles, interference from neighboring radio and electrical systems). By proper planning of the radio access network, it may be possible to minimize the potential signal loss over the radio links—especially those caused by multipath fading effects. Most WLL systems require LOS radio links to ensure adequate radio range and will deploy one or more base stations (cell sites) to provide adequate radio coverage for their subscribers.

In order to maximize the probability that LOS paths to the base station will be available from all of the subscriber terminal locations to be served by the base station, radio planning needs to be carried out on a cell-by-cell basis. This requires close attention in the selection of base station site and antenna height as well as the antenna height at subscriber terminal locations. For WLL systems there is relatively much greater flexibility in the choice of base station locations compared to the location of radio subscriber outdoor units and antennas.

In principle, it is necessary to choose a base station site that provides good visibility to the desired coverage area of the base station. At the time of base station site selection the exact locations of the remote subscribers is generally unknown. The planner only has information on the broad geographic service area where the subscribers are located and radio coverage is required. It is therefore not possible to carry out a definitive LOS check between the base station and subscriber units—only a general LOS coverage check is possible. For most WLL systems, such location planning can be carried out by checking for the following:

1. There are no obvious obstacles between the base station and the intended geographic service area that may contribute to multipath fading.

2. There are no other base stations in the neighborhood that may lead to mutual interference.
3. The height of the base station antenna extends over surrounding buildings and terrain variations to avoid shadowing effects.
4. General terrain surrounding the base station site is sloping down to ensure maximum visibility for the base station.

Besides the fact that the base station needs to be located where it can efficiently and economically provide coverage for the potential subscribers, the exact location of the base station may also be constrained by such factors as site availability, cost of leasing the site, and local regulations.

Similarly, some basic LOS confirmation procedures can be employed during the installation of subscriber terminal equipment. These are especially required for WLL applications in rural environments to maximize system range and coverage. In the case of short- to medium-length radio links that are less than 10 km, availability of LOS between the remote subscriber location and the serving base station can be confirmed visually (using binoculars where necessary).

In case the visual LOS check indicates the presence of an obstacle, it may be necessary to determine the height of the antenna mast in order to overcome the blockage caused by an obstacle. This may be achieved by sending up a helium balloon at the remote site and observing it visually from the base station antenna location.

Additionally, tests may be performed to check that the Fresnel zone between the remote subscriber location and the base station is clear of obstacles and that the area near the intended location of the antenna is clear of any reflecting surfaces.

If an LOS path is not possible between the remote station antenna and the intended serving base station, it may be necessary to seek an LOS path to an alternative base station—if such an option is available.

Whereas good planning of base station locations and suitable antenna heights and placing will go a long way toward ensuring LOS paths and avoiding multipath fading effects, additional counter measures like adaptive equalization and/or space diversity can also be employed. An adaptive equalizer corrects any amplitude, frequency, or phase distortion introduced on the radio path and attempts to restore the signal to its original balance. The equalization function needs to be implemented at the front end of the receiver before the demodulation stage.

In the case of space diversity, two separate receiving antennas are deployed with a suitable mixing function in the receiver to instantaneously provide the strongest signal. CDMA radio technologies are inherently well-suited to provide this type of space diversity, and this feature of CDMA-based WLL systems is a distinct advantage over systems based on alternate radio technologies. In fact, existence of multipath fading may be an advantage in CDMA systems in that CDMA systems are able to select the best incoming signal.

5.7 RADIO COVERAGE AND FREQUENCY PLANNING FOR WLL SYSTEMS

As described in Chapter 3, a large geographic area to be served by a cellular mobile system is divided into cells with diameters from 2 to 50 km, each of which is allocated a number of radio-frequency (RF) channels. Transmitters in each adjacent cell operate on different frequencies to avoid interference. However, transmit power and antenna height in each cell is relatively low so that cells that are sufficiently far apart can reuse the same set of frequencies without causing co-channel interference. The same principle of a hexagonal grid and frequency reuse is deployed for planning of WLL systems to provide radio coverage for its subscribers.

As the demand for WLL service grows, additional cells can be added, and as traffic demand grows in a given area, cells can be split to accommodate the additional traffic. For radio planning purposes the potential coverage area is divided into cells in a regular fashion. Typical cell clusters used in WLL radio coverage planning are illustrated in Figure 5.3. Cluster sizes of $N = 1, 3, 4, 7$, and 12 are commonly used.

An example of frequency reuse in a seven-cell cluster is shown in Figure 5.4. A particular group of channels F1 is used in one cell which is then reused in another cell with the same coverage radius at a distance D which is given by

$D = R_c(3N)^{1/2}$, where R_c is the cell radius and N is the reusability factor (=7 in this case)

The planning of radio coverage for WLL systems is somewhat different than for cellular mobile systems in terms of where and when the coverage needs to be provided. As illustrated in Figure 5.5, the coverage for cellular mobile systems needs to be provided over the geographic service area where the subscribers are expected to travel

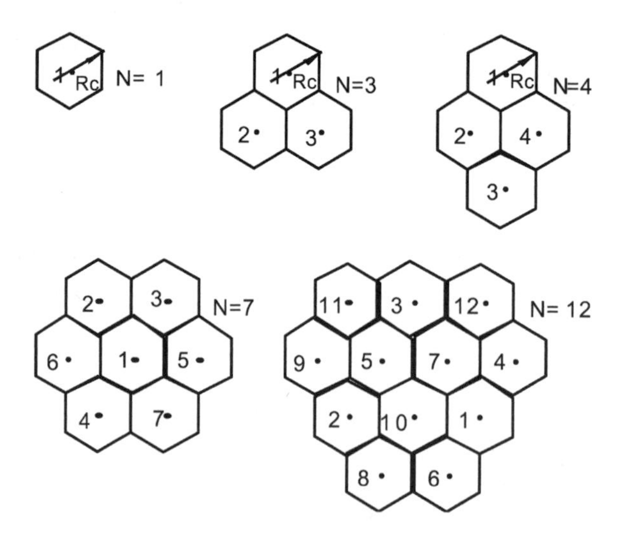

Figure 5.3. Commonly used cell clusters for radio coverage planning.

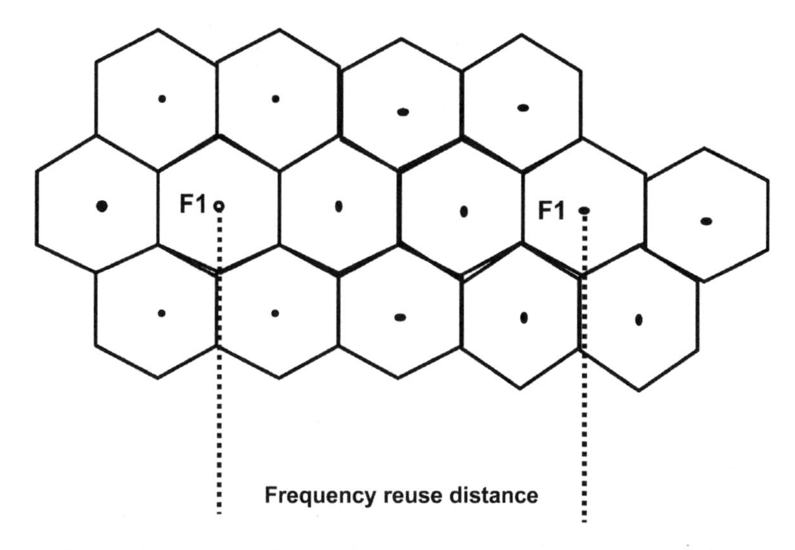

Figure 5.4. Seven-cell cluster and associated frequency reuse distance.

and use the service. For example, coverage is required along all major transportation facilities. Furthermore, the infrastructure to provide the coverage over the entire service area needs to be available right from the beginning of service deployment.

For WLL systems on the other hand (with no mobility), the radio coverage is required only in those specific areas where the subscribers reside and the

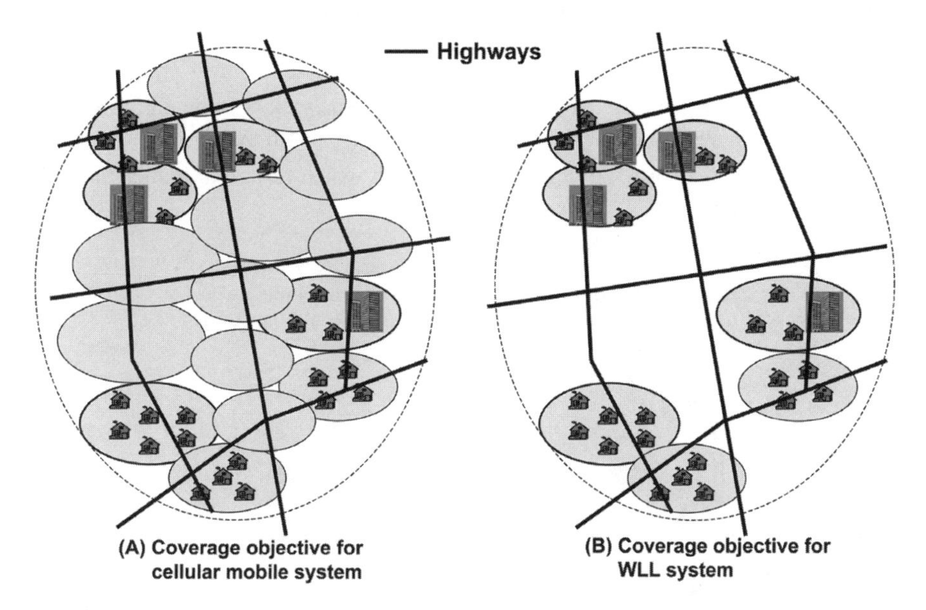

Figure 5.5. Radio coverage requirements for cellular mobile versus WLL systems.

infrastructure for providing the coverage can be gradually phased in as the WLL subscriber base grows in volume. Furthermore, the requirement to provide continuous coverage over very large areas may be less prevalent in WLL networks than in cellular mobile networks. For cellular mobile networks, frequency reuse and associated planning is almost always required. For WLL systems the following two coverage conditions are generally encountered:

- Broad multicell continuous coverage—involving extensive frequency reuse within the coverage area. This condition may be representative of high-capacity networks in urban or metropolitan environments where a large number of small cells are employed.
- Localized, selective coverage with a single or small number of cells which represents coverage for small *islands* such as key population centers (villages, towns) with areas of limited or no service in between these islands.

5.8 SECTORIZED CELLS AND FREQUENCY REUSE PLANNING FOR WLL SYSTEMS

Maximizing the capacity of the base station (i.e., the ability of the base station to meet all the bandwidth demands of the remote subscriber stations) in a WLL system is a problem that is frequently encountered by WLL system planners. Furthermore, the system planner also needs to address the range and coverage of the system that may be affected by unavoidable obstacles in many of the links.

As shown in Figure 5.6, though omnidirectional antennas can be used at the base station, in current WLL systems it is more common to use *sectorized* antennas because they provide the following benefits:

- The same frequencies can be reused in different sectors allowing an overall increase in system capacity (specially in CDMA-based systems).
- Sector antennas provide higher antenna gain which results in higher range or better availability.

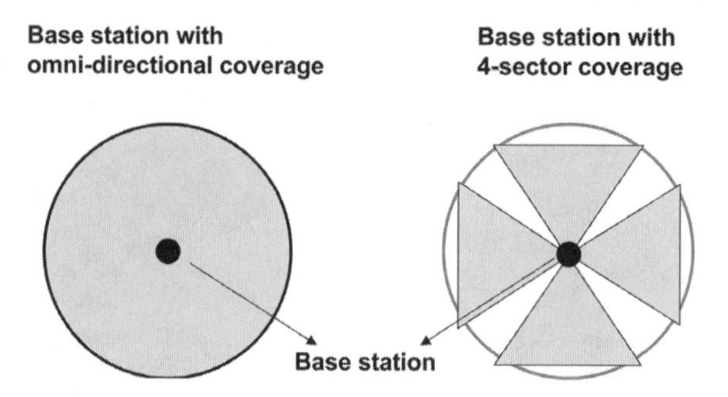

Figure 5.6. Use of sector antennas at WLL base stations.

- Sector antennas provide "positioning" flexibility to overcome obstacles and obtain optimum coverage and capacity.

The number of sectors deployed at the base station site can be 2, 3, 4, or more sectors depending on the prevailing needs for coverage and range. However, the cost of the base station will increase with the number of sectors due to the number of directional antennas required and the increased complexity of the transceiver systems. Each sector is allocated its own group of frequency channels based on the frequency reuse pattern deployed for maximum spectral efficiency. In a WLL system, deploying a large number of base stations, four-sector base stations can be used with a frequency reuse pattern of two or four. The main consideration in choosing a frequency reuse plan where sectorized antennas are deployed is to minimize interference between adjacent sectors within the same base station and/or interference between sectors from neighboring base stations. Both co-channel and adjacent-channel interference can affect C/I ratio at the receiver.

A system with four sector base stations and a frequency reuse pattern of two is illustrated in Figure 5.7. All allocated frequencies for the WLL system can be reused at least once in each cell; that is, the available frequency channels are divided into two groups (1 and 2), each group being used in sectors that are diagonal to each other at a base station. The same set of frequency channels can then be used by adjacent base stations, resulting in a frequency reuse factor of 2.

The use of this frequency plan with two-frequency reuse leads to potential interference between adjacent channels along adjoining sectors mainly caused by unavoidable overlap in the spectrum masks of adjacent channels and nonideal nature of practical antenna patterns. The following steps are generally taken to mitigate this adjacent-channel interference problem:

- Group of frequencies allocated to adjacent sectors are generally staggered. Using the example of GSM channel structure (consisting of 124 full duplex

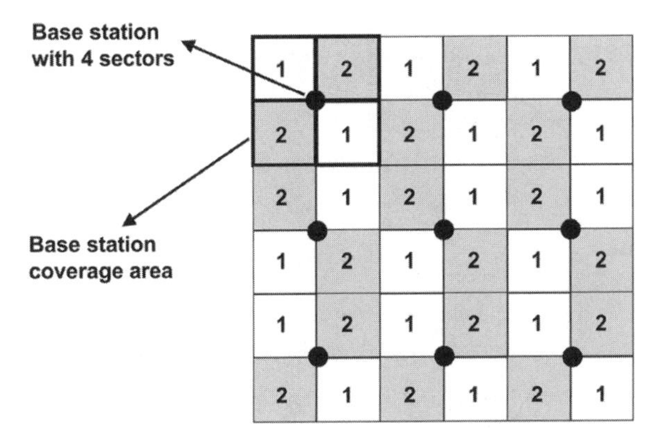

Figure 5.7. Two-frequency reuse pattern with four sector BS antennas.

channels), group 1 will therefore include even-numbered channels (0, 2, 4,..., 124) and group 2 will consist of odd-numbered channels (1, 3, 5,..., 123).

- The frequency reuse pattern is reversed in appropriate adjacent base stations. For example, frequency groups allocated to the four sectors for base stations in row one and row two in Figure 5.7 are reversed—and similarly between base stations in row two and row three.
- If possible, redirecting a remote subscriber antenna to an alternate base station (rather than its regular home base station) for remote subscriber locations encountering adjacent-channel interference.
- Use antennas with good side-lobe and back-lobe suppression characteristics at remote subscriber locations.

If a sufficient number of radio channels are available to meet the traffic demand, a four-frequency reuse plan in a four-sector base station arrangement is preferred over a two-frequency plan. Such a plan, illustrated in Figure 5.8, generally provides better interference reduction characteristics, when interference reduction arrangements similar to those described above for two-frequency reuse are deployed.

Again using the GSM example, in this frequency plan the 124 GSM channels (0, 1, 2, 3,..., 123) will be arranged as follows for each of the four sectors:

Sector 1: channel numbers 0, 2, 4, 6,..., 60
Sector 2: channel numbers 62, 64, 66, 68,..., 122
Sector 3: channel numbers 1, 3, 5, 7,..., 61
Sector 4: channel numbers 63, 65, 67, 69,..., 123

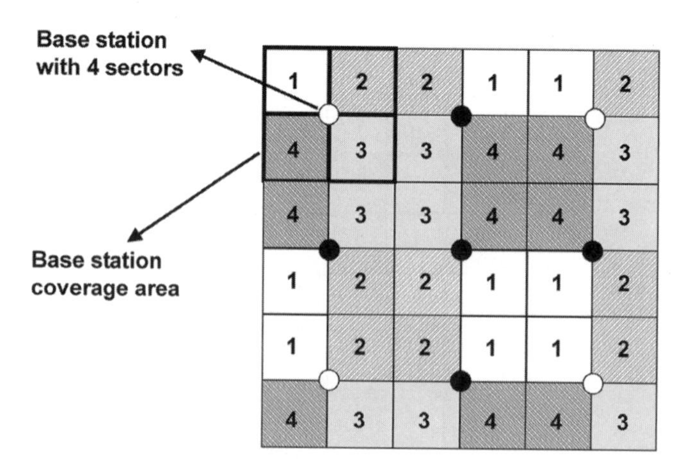

Figure 5.8. Four-frequency reuse pattern with four sector BS antennas.

This arrangement ensures maximum separation between frequencies used in the adjacent sectors in a four-sector base station.

Furthermore, as illustrated in Figure 5.8, the frequency group allocations to individual sectors in base stations in row two are rotated by 90 degrees as compared to base stations in row one (and back into their original position in row three). This provides maximum potential for redirecting remote subscriber antennas to alternate base stations, if such redirection is required to minimize interference at specific subscriber locations.

5.9 CAPACITY ENHANCEMENT METHODS: CELL SECTORIZATION AND CELL/SECTOR SPLITTING

Because of the inherent frequency reuse capability of a cellular system, at least theoretically, the coverage and capacity of a cellular system can be increased indefinitely by adding more cells and reducing the cell size, respectively. However, there are practical limitations imposed by need for more frequent handoffs (where mobility is to be supported) and increased interference when cell sizes become very small and cell densities become large within a geographic coverage area. In such a case, allocation of additional frequency spectrum becomes the only option for increasing capacity.

For economic reasons, cellular mobile as well as WLL systems are planned to evolve with increasing subscriber base and traffic demand. The initial WLL radio network may be based either on (a) a cellular pattern with omnidirectional antennas (Figure 5.3) located at the center of the cells or (b) using a sectorized antenna plan (Figures 5.7 and 5.8). In the case of an initial cellular plan, additional capacity can be provided either by (a) replacing the omnidirectional antennas at the cell centers by 120-degree antennas and reorganizing the cells so that the antenna positions appear at appropriate vertices of the hexagonal cells or (b) replacing the omnidirectional antennas by 120-degree or 60-degree sector antennas without changing the antenna location with respect to the original cell sites.

These two options are illustrated in Figure 5.9. It should be emphasized that in Figure 5.9(B), the actual location of the cell sites (i.e., location of the transceiver stations) remains the same; it is the cell boundaries that are reconfigured so that the antenna sites now appear at the appropriate vertices of the hexagonal coverage areas. Generally, changing existing cell sites is not economical because it involves new site acquisition and antenna mast construction.

The concept of *cell splitting* is illustrated in Figure 5.10, whereby some of the cells may be split in two or more cells in order to accommodate additional demand. Sometimes smaller cells are overlaid on larger cells to provide sufficient coverage in congested or high interference areas like large metropolitan business centers.

In the case where the initial WLL system network is based on sectorized antennas for frequency reuse (Figures 5.7 and 5.8), capacity increase can be achieved by replacing one or more original sectors by two (or more) new sectors (sector splitting). Two examples of such sector splitting are illustrated in Figure 5.11 for an original four-sector two-frequency reuse plan that was shown in Figure 5.7.

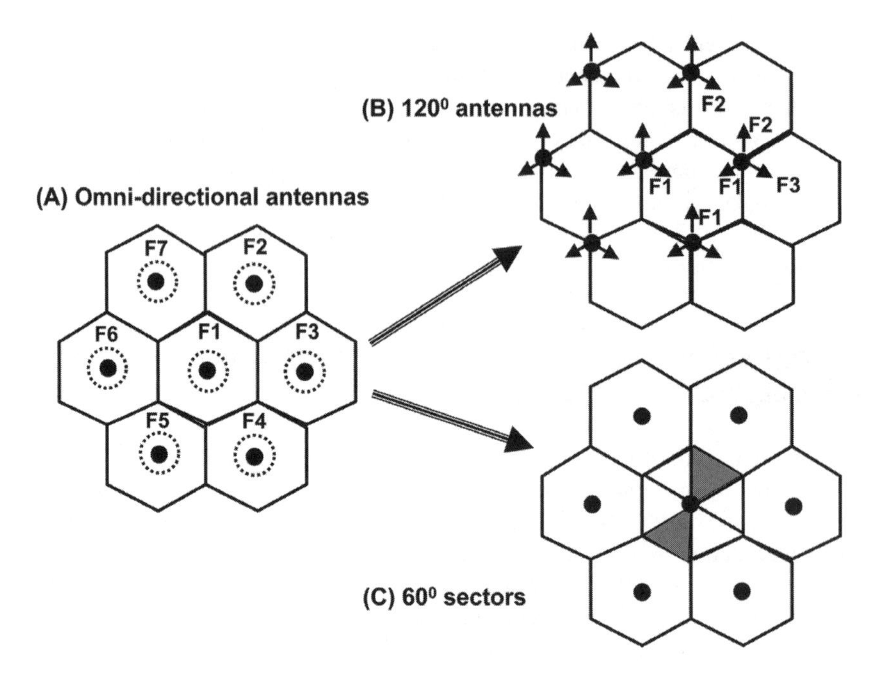

Figure 5.9. Replacement of omnidirectional antennas by sector antennas to increase system capacity.

In Figure 5.11(B), each upper 90-degree adjacent sector has been split into two 45-degree sectors with appropriate frequency assignments to minimize interference. In Figure 5.11(C) the two diagonal 90-degree sectors are split into four 45-degree sectors. However, in this case, original frequency assignment to the upper left sector has been changed to provide better interference suppression.

In order to improve on the co-channel discrimination and/or to overcome interference caused by an obstacle in the radio path, sometimes use is made of cross-polarization discrimination properties of the directional antennas at the base station and the remote subscriber stations. An example of a four-sector, two-frequency

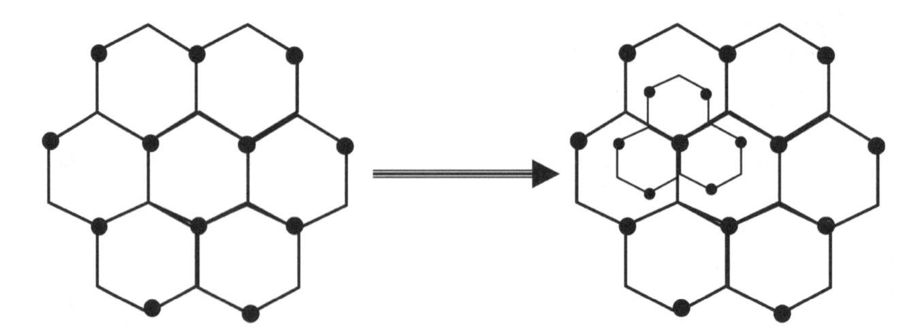

Figure 5.10. Cell splitting to increase system capacity.

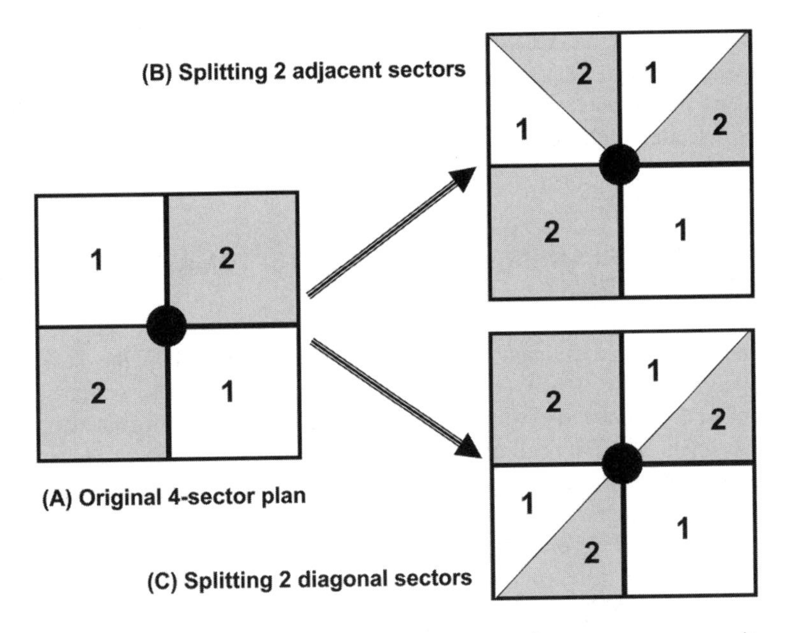

Figure 5.11. Examples of sector splitting to increase system capacity.

reuse plan using vertical and horizontal polarizations is illustrated in Figure 5.12. The figure also illustrates how the four 90-degree sectors may be split into six 60-degree sectors to increase system capacity.

In practice, the use of cross-polarization discrimination is generally avoided because of the following reasons:

- Horizontally polarized radio signals are more susceptible to loss due to rain and other precipitation, thereby adversely affecting the system range.

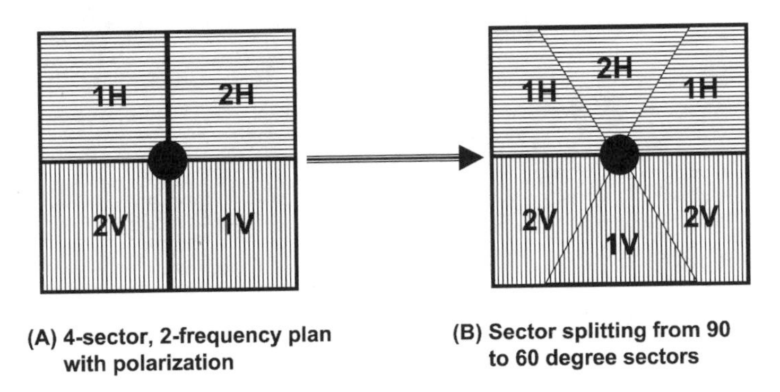

Figure 5.12. Examples of deploying antenna polarization.

- Any changes in antenna orientation at the base station (e.g., when new sectors are created through sector splitting) may also require reorientation of the antenna at the remote subscriber location—which can only be achieved by sending a technician to physically rotate the antenna at the subscriber site.

5.10 FREQUENCY PLANNING REQUIREMENTS: TDMA- VERSUS CDMA-BASED SYSTEMS

The frequency reuse and frequency planning aspects discussed in the last few sections largely apply to TDMA (or FDMA)-based systems. In these systems, the fixed amount of frequency spectrum allocated to the system is partitioned into a number of channels, and each cell or sector is allocated a fixed number of channels based on the frequency reuse plan and the traffic needs. This requires considerable work to layout the frequency planning for a network, and it becomes more complex as cells are split and the network grows. It should be emphasized that due to physical limitations, in practice, cells are not regular hexagonal shapes and they are not spaced out in a nice orderly fashion.

However, as shown in Figure 5.13, for a CDMA-based system the same frequency channels are assigned to neighboring cells, and no complex frequency planning is required. For CDMA-based systems, each cell and sector is assigned a different long code (PN code) which allows the mobile to discriminate between the signals from the serving BTS and those from neighboring BTS and avoid interference. CDMA-based systems therefore may not require frequency planning per se, but they will require planning for assignment of PN codes to individual cells and/or sectors. Some CDMA-based WLL systems recommend the use of two-frequency reuse plan, thereby requiring some (though minimal) frequency planning.

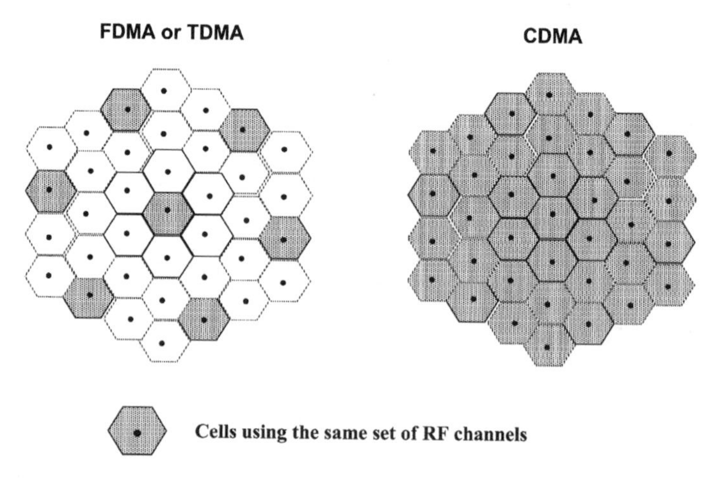

Figure 5.13. Frequency planning and reuse for FDMA or TDMA versus CDMA systems.

For systems consisting of a large number of cells/sectors where extensive frequency planning is called for, some type of commercially available software planning tool is deployed. These tools can quickly generate multiple, alternate frequency plans so that the system layout can be optimized for capital and operating costs. In other words, frequency planning requirements should not play a major part in the choice between a TDMA- or CDMA-based WLL system.

5.11 COMPARISON OF RADIO PLANNING ASPECTS FOR WLL VERSUS CELLULAR MOBILE SYSTEMS

This chapter has attempted to cover the radio characteristics and radio planning aspects for WLL systems. These are summarized in Table 5.2, which also contains a comparison with cellular mobile systems for these radio planning aspects.

TABLE 5.2. Radio Planning for WLL Systems: Comparison to Cellular Mobile Systems

Radio Aspect	WLL Systems	Cellular Mobile Systems
Frequency bands	No dedicated bands. Sharing with other applications may be required.	Dedicated bands that are highly regulated.
Radio propagation environment	Generally, line-of-sight paths subject to multipath fading.	Indirect paths subject to Rayleigh fading.
Path loss and fade margins	Mainly multipath and temporal fading margins. No margins for in-building and roaming aspects.[a]	Up to 40-dB margin for mobility. Fading effects built into receiver sensitivity.
Radio coverage requirements	Often noncontiguous islands. Coverage needed in areas where subscribers reside.	Generally contiguous over large areas. Coverage needed to support user mobility.
Frequency reuse patterns	Typically, omni cells with reuse of seven- or four-sector with reuse of two.[b]	Typically, omni cells with reuse of 7 or 12, or three-sector cells with reuse of 4[b].
Subscriber antenna characteristics	Directional antennas requiring proper orientation and height to ensure line-of-sight path.	Omnidirectional antennas for use in multiple environments (indoors, in vehicles).
Subscriber growth and expansion	Needs to be factored in initial plan to avoid large-scale reorientation of subscriber antennas if cells are split or sectorized.	Cell additions, cell splitting, cell sectorization generally used to add capacity and/or coverage.

[a]Assumes a fixed subscriber antenna.
[b]May not apply for CDMA-based systems.

CHAPTER 6

Planning and Deployment Aspects of WLL Systems

This chapter attempts to provide a broad understanding of the planning process that may be required for the design and installation of a WLL system. The various steps starting from assessing the service needs of the potential subscribers to the planning of the radio components and their interconnections are described. A very brief introduction to the business aspects in the form of preparation of a business case for a WLL system required for assessing the economic viability of the enterprise is also included.

6.1 INTRODUCTION

Planning and deployment of a WLL system involves a number of steps, some of which may be interrelated, thus requiring a number of passes through these steps. The decisions that the planner finally makes may not completely depend on deterministic data. One needs to recognize that the inputs to the planning process are not limited to technical factors. Many other regulatory and socioeconomic factors are also likely to play a role in the decision process. The key steps involved in the deployment planning of WLL systems include

- Assessing the service needs of the intended customer base, including
  - Effects of supporting mobility
  - Future service requirements and associated capabilities
- Estimating the traffic load and its distribution
- Choosing a suitable WLL technology
- Choosing an economic and efficient plan for the radio system
- Choosing appropriate transport and protocol options for network interconnection

In most instances, the process will also include the development of a business model and a business case in order to ensure that the selected WLL system design will lead to an economically viable and profitable enterprise.

Introduction to WLLs. By Raj Pandya
ISBN 0-471-45132-0

6.2 ASSESSING THE SERVICE NEEDS

Depending on the subscriber environment (residential, business, or mixed) to be served by the WLL system, an assessment needs to be made on the set of services that the system may need to support—both at initial deployment and in the foreseeable future. Following is a list of potential services that may need to be considered:

- PSTN (local calls only)
- PSTN (local + long-distance)
- ISDN (basic rate)
- Fax
- Data
- Supplementary services
- Centrex
- Multiple lines
- Leased lines
- Internet services
- Mobility services

PSTN (Local Calls Only): This is the basic voice service, which all WLL systems provide as a minimum offering. The key concern here is voice quality. In most markets, voice quality equivalent to wireline systems will be essential.

PSTN (Local + Long-Distance): If the WLL operator owns the PSTN switch, then addition of long-distance service is straightforward. However, WLL operators who only own the radio infrastructure may have to pay significantly higher interconnection charges to the local phone company if it is to switch and carry long-distance traffic.

ISDN (Basic Rate): This provides two 64-Kb/s lines, either of which can be used for voice or data traffic. It also provides a 16-Kb/s path for signaling.

FAX: Group 3 (analog) and/or Group 4 (digital) fax will generally be available in most WLL systems.

Data: The key issue here is the data rate that needs to be supported. Analog lines can typically support modem data rates of 9.6 kb/s. Digital lines can support much higher rates (64^+ kb/s).

Supplementary Services: These include such features as caller ID, call forward, call waiting, advice of charge, and so on. Many of these services can be provided on non-ISDN digital switches.

Centrex: This is a virtual PBX service provided by the local switch. It means that the employees of the organization having the Centrex facility can communicate with each other using numbers with reduced digits similar to a PBX. However, these internal calls have to be switched at the PSTN switch, thereby adding load on the radio interface.

Multiple Lines: Businesses generally require multiple lines ranging from 2 to 10, and sometimes more. Systems capable of providing this capability are available, but are configured for business customers only. Off-the-shelf WLL systems are not designed to support multiple lines.

Leased Lines: Option is difficult to support in a WLL system where dedicated resources over the radio interface are not feasible. This service therefore is not generally available on WLL systems.

Internet Services: A WLL system which can provide a line that can support a reasonable data rate (either through a modem or as a digital line) can be used to access the Internet. It is also possible for the WLL operator to become an ISP by installing router capabilities.

Mobility Services: In some instances, limited mobility may be provided to the WLL subscribers. Provision of mobility in a WLL system raises certain special issues related to performance and cost. These are discussed in the next section.

While assessing the service needs in terms of one or more of the above services, the planner also needs to take into account such future scenarios as: evolving service needs, changes in the size of the customer base, and expansion in the coverage area.

Evolving Service Needs: It is important to make appropriate provision in the deployment planning process to ensure that future service requirements can be met. Services considered may not be required initially, but may be needed in the future. In fact, other functions and attributes, not now contemplated, and sometimes not yet known, may become a future requirement. The network deployment plan and the selected technology need to be sufficiently flexible and sufficiently capable to respond effectively and economically to future service needs.

Customer Base: The number of customers served will undoubtedly grow over time. The network deployment plan must be able to increase capacity in a convenient and economic manner, so as to provide good continuing service as the customer base grows.

Coverage Area: It may well be appropriate in the future to expand the coverage area in which service needs to be provided. The network deployment plan should be prepared to ensure that expansion opportunities can be implemented in an orderly and economic way.

Some of the approaches for capacity and coverage expansion using such techniques as cell sectorization and cell and sector splitting were addressed in Chapter 5.

6.3 SUPPORT OF MOBILITY IN WLL SYSTEMS

The provision of mobility has the potential to add significantly to the complexity and cost of a WLL system, due to enhanced coverage needed and the additional network elements that have to be provided. In order to support full mobility, cordless and proprietary WLL systems require significant enhancements in coverage and additional equipment over and above that for standard WLL service. The levels of mobility provided by a WLL system may be classified as follows:

- *Full mobility* is the mobility offered by cellular radio, in that the user can expect roaming facilities over a wide area and intercell handoffs during call in progress.
- *Limited mobility* is the case where the user is able to move within a limited area near the home, perhaps in a radius of a few kilometers, and still make and receive calls, with handoff facility provided (but no roaming).
- *No mobility* is the case where the subscriber has a telephone that is attached to a cord or has a cordless set for use within the premises.

Another level of mobility that is sometimes mentioned is the ability of a subscriber terminal or customer premises equipment to relocate in non-real time (generally referred to as *nomadicity*). The case of nomadic terminals can be supported by WLL systems that support either full or limited mobility.

The underlying effects and requirements to support different levels of mobility are captured in Table 6.1.

Mobile users expect that they will be able to receive and make calls from anywhere including from inside buildings. There is a very significant loss in signal strength (up to 20 dB) for signals to/from terminals inside a building. Furthermore,

TABLE 6.1. Effects of and Requirements for Supporting Mobility on WLL Systems

Effects/Requirements	Full Mobility	Limited Mobility	No Mobility
Signal loss: building penetration	Yes	Yes	No
Signal loss: omni subscriber antennas	Yes	Yes	No
Available level of link budget[a]	Low	Low	High
Cell site coverage area[a]	Low	Low	High
Subscriber location capability	Yes	No	No
Subscriber authentication capability	Yes	No	No
Wide area roaming capability	Yes	No	No
Intercell handover capability	Yes	No	No
Complex network planning/operation	Yes	No	No

[a]Depends on deployed technology and frequency of operation.

mobile terminals have to use omnidirectional antennas to receive and transmit signals from different locations relative to the base station antenna. Compared to standard WLL system (i.e., no mobility) where typically directional antennas are deployed at the subscriber location, an additional loss in signal strength of up to 20 dB may be incurred. This requirement for up to an additional 40 dB in link budget therefore drastically reduces the size of the cells that may need to be deployed with concurrent cost increases to service a given coverage area.

In the case of WLL systems based on cordless technologies like DECT and PHS, or proprietary technologies, which are not designed for wide area mobility, the option of providing full mobility is really not viable. Use of propagation models applicable to DECT suggests that a 40-dB difference in link budget would reduce a cell size of 5 km without mobility, to less than 400 m with mobility—which will translate into 600 times more cells (hence cost) if full mobility is to be offered over the same coverage area (W. Webb, *Introduction to Wireless Local Loop*, second edition, Artech House, 2000).

In the case of WLL systems based on cellular technologies, where the WLL terminals are typically mounted inside the house, this is not an important issue. The 40-dB link budget is already built into a cellular system, and the range of a cellular-based WLL system is the same as that of the cellular mobile system. Mobility is therefore a viable option for WLL systems based on cellular technologies.

In order to support *full mobility* in a WLL system, additional capabilities indicated in Table 6.1, which are standard in a cellular system, will be required. These capabilities in cellular systems lead to the need for such network elements as HLRs, VLRs, authentication centers, and gateway mobile switching centers, along with capabilities for switching calls between base stations and between MSCs during the progress of a call as the subscriber moves across cell boundaries. With the increased number of cell sites and network elements to support full mobility, the WLL system will require more complex radio and network planning as well as operational support.

In the case of *limited mobility*, many of these capabilities are not needed, because the terminal is allowed to move in a limited space (e.g., within the coverage area of one or a few base stations) and a lower level of authentication and security environment may exist. Many WLL systems currently deployed in developing countries provide limited mobility as an option at a much lower cost and therefore compete with existing cellular mobile networks.

6.4 ESTIMATING TRAFFIC DENSITIES AND TRAFFIC DISTRIBUTION

Having determined the range of services to be supported by the intended WLL system, it is then necessary to estimate the nature and extent of the traffic that will be generated by the potential subscribers of the WLL system. This forms the primary input for estimating the number and size of the cells, as well as the frequency planning for the cells. The factors that need to be considered include

- Subscriber densities
- Subscriber growth estimates
- Subscriber distribution
- Subscriber types
- Subscribers' usage patterns
  - Type of services invoked
  - Level of usage (erlangs, bits)
  - Busy hour

Subscriber density is a key factor in planning and deployment of a WLL system. Note that this density addresses *subscribers* to the system and not merely the population density in the target area.

Subscriber growth or growth in the subscriber base must also be considered, since not all potential customers will sign up for the service initially. Experience indicates that the subscriber base will increase over time, and the inherent difficulty in predicting where and when this will occur needs to be recognized.

Subscriber distribution must be considered within the concept of subscriber density; the subscriber distribution is also very important in estimating the service requirements. It is necessary to know if the potential customer base is evenly distributed over the target area or, as is more typically the case, where and how the potential subscribers are clustered.

Subscriber types generally determine the type and volume of traffic that would have to be supported. Subscribers may be classified as residential or business, or rural, suburban, urban, and metropolitan.

Subscriber usage patterns are essentially used to estimate traffic/subscriber and the overall busy hour traffic load for the WLL system design.

Whereas the estimates for subscriber densities, subscriber distributions, and subscriber types can be estimated based on information on demographics and suitable surveys, subscriber usage can be estimated using standard traffic usage tables. For example, ITU-T Recommendation Q.543 provides reference traffic loads expected from different subscriber types (residential, business etc.) in terms of busy hour erlangs and busy hour call attempts (BHCAs). These figures can be utilized as the basis for estimating the overall traffic load to be supported by the WLL system.

6.5 SELECTION OF A SUITABLE WLL TECHNOLOGY—INFLUENCING FACTORS

The main factors that need to be considered in deciding on a suitable technology for the WLL system include

- Ability to operate in the available frequency spectrum
- Subscriber density and target market to be served
- System capacity and range
- Capital and operational cost
- System functionality to meet target service needs

The choice of a suitable technology, once the list is narrowed down to those which will operate in the available frequency band, is primarily dictated by the capacity, cost, and the functionality the system can support and its suitability for the target market (rural, suburban, urban, metropolitan, etc.).

The above factors essentially focus on achieving conformity with relevant technical specifications and optimization of capital and operational costs of the equipment. However, in making a suitable choice between WLL technologies, for either a new network or an existing network, a range of secondary factors may also need to be considered. For example, one or more of the following factors may affect the ultimate choice for the technology:

- Maturity and track record of the technology
- Technology transfer considerations
- Availability of skilled staff for installation and maintenance
- Ongoing technical and operational support
- Suitability of the technology for multivendor options
- Flexibility of the technology in accommodating changes or errors in planning
- Future-proofing aspects of the technology

A further crucial factor is the need to ensure that equipment selection is undertaken within the framework of the operator's strategic, long-range network development plan. This will ensure the strategic development of the network and should limit, for example, problems of interworking between different parts of the same network.

6.5.1 Ability to Operate in a Given Frequency Band

WLL systems are available in a range of frequency bands. Some of these frequency bands may already be in use for other applications or services (e.g., GSM cellular systems in the 900-MHz and 1800-MHz bands). This implies that if the WLL and a cellular system will be operating in the same frequency band and in proximity to each other, radio engineering for the WLL system needs to be carefully planned to avoid interference between the signals from the two systems. It should be noted that CDMA-based systems lead to least interference, when operating in a region where the same spectrum is shared between more than one service or between more than one operator. Table 6.2 illustrates the range of technologies for WLL systems

TABLE 6.2. Examples of WLL Technologies Available in Different Frequency Bands

Frequency Band	Available WLL Technologies
450 MHz	Analog cellular (NMT)
800 MHz	Analog cellular (AMPS, NMT, TACS)
	Digital cellular (DAMPS, cdmaOne)
900 MHz	Digital cellular (GSM)
1.8–2.0 GHz	Cordless telecommunications (DECT, PHS)
	Digital cellular, PCS (GSM, DAMPS, cdmaOne, PACS)
2.0–2.5 GHz	Proprietary (Airspan)
3.4–3.6 GHz	Proprietary (Nortel: Internet FWA, Lucent: AirLoop)
>10.0 GHz	Broadband wireless access (LMDS, MVDS, IEEE802.16, ETSI HIPERACCESS, SATCOMs)

available in different frequency bands. Note that a given WLL technology may be available in multiple frequency bands.

Systems that are based on cordless standards (like DECT and PHS) or cellular standards (like GSM and cdmaOne) are vendor-specific systems that utilize the air interface specifications and the frequency bands standardized for the relevant cordless or cellular system. The so-called *proprietary* systems define their own air interface specifications and make their own choice for the frequency band of operation (e.g., Nortel Networks' Internet FWA system has its own air interface design using TDMA/FDD technique operating in the 3.4 to 3.6-GHz band).

6.5.2 Subscriber Density and Target Market

The subscriber density environments can generally be classified as

- Very low density rural
- Rural
- Suburban
- Urban
- Metropolitan

For very low density and low-density market environments, analog cellular technologies (NMT450, TACS, AMPS) or digital cellular technologies (GSM, DAMPS, cdmaOne) can be deployed. For urban and metropolitan environments, cordless technologies (DECT, PHS) as well as digital cellular technologies (GSM, DAMPS, cdmaOne) are suitable. Almost all technologies are deployable in suburban areas. Proprietary WLL technologies are available for all but the very low subscriber density situation. It is interesting to note that cdmaOne technology can be suitably adopted for all levels of subscriber densities.

6.5.3 System Capacity and Range

Capacity of a WLL or cellular system is typically quoted in terms of the maximum number of voice channels per cell per MHz of bandwidth or sometimes in terms of the maximum data throughput (kb/s) that the data channels in a cell will support for the allocated spectrum. Typical system capacity and system range associated with various WLL technologies are illustrated in Table 6.3. The information in this table is gleaned from various vendors' literature and should be used for preliminary planning purposes only.

The capacity figures provided in Table 6.3 are not absolute and should be used for comparative planning only. Furthermore, cordless technologies like DECT and PHS provide relatively few channels per cell but have inexpensive and easy-to-deploy base stations that can be used with a short range. By deploying more base stations in a larger area, a DECT network could potentially provide higher capacity than some proprietary WLL systems at comparable cost.

6.5.4 Capital and Operational Cost

Estimation of cost for a WLL system requires an understanding of the range and the capacity of different systems. In the calculation of system cost, a key parameter is the number of base stations required, because the number and their interconnection architecture will determine the sizing of various other network elements. Thus, a first cut estimate of the relative costs of different WLL technologies can be obtained by estimating for each technology the number of base stations required (for current and forecasted coverage) and multiplying it by the cost per base station. The success of this rather simple approach depends on the accuracy of the available figures for base station costs.

Cordless technologies generally have the lowest cost for base station equipment (about half of other technologies). However, since they have a limited range,

TABLE 6.3. Nominal System Capacity and Range for Some WLL Technologies

WLL Technology	Capacity (Voice Channels/Cell/MHz)	Maximum Range (km)	Multiple Access Method
Analog cellular	3.3	50	FDMA/FDD
DECT cordless	5.2	5	TDMA/TDD
PHS cordless	8.0	5	TDMA/TDD
GSM cellular	10.0	30+	TDMA/FDD
cdmaOne cellular	24.0	40+	CDMA/FDD
Proprietary			
Nortel Internet	11.0	40+	TDMA/FDD
Lucent AirLoop	8.5	30+	CDMA/FDD

TABLE 6.4. Functionalities and Capabilities Supported by Various WLL Technologies

WLL Technology	Voice Quality	Max Data Rate (kb/s)	ISDN	Fax	Multiple Lines	SS Range	Mobility Support
Analog cellular	Poor	1.2	No	No	No	Limited	Yes
DECT cordless	Fair	4.8–550	Yes	Yes	Yes	Wide	Limited
PHS cordless	Fair	4.8–32	No	No	No	Limited	Limited
GSM cellular	Fair	9.6–128[a]	No	Yes	No	Wide	Yes
cdmaOne cellular	Good	9.6–64	No	Yes	No	Wide	Yes
Proprietary							
Nortel Internet	Good	192	Yes	Yes	Yes	Wide	No
Lucent AirLoop	Good	128	Yes	Yes	Yes	Wide	No

[a]Assumes GPRS enhancement.

additional base stations may be required to provide adequate coverage, thereby increasing the cost of the system.

6.5.5 System Functionality to Meet Target Service Needs

The functionality and capability of a WLL technology to support a range of services also needs to be assessed in order to ensure that the quality and range of services expected by the subscribers will be met by the selected system. Table 6.4 illustrates the range of functionalities and capabilities supported by major WLL technologies.

6.6 SOME ADVANTAGES OF CDMA TECHNOLOGY FOR WLL SYSTEMS

As discussed in the previous section, there are multiple factors that need to be considered for selecting a suitable WLL technology. However, CDMA technology provides the following advantages when used in WLL systems:

- It has higher capacity (in terms of subscribers/km^2/MHz).

- It exhibits significant capacity gain through sectorization because the same frequency can be used in each sector leading to a high capacity/cost ratio for sectorization.
- No frequency planning is needed because all cells/sectors use same frequency allocation.
- Space diversity available in CDMA systems provides immunity from multipath fading and capability for soft handoff (where mobility is supported).
- CDMA systems are capacity-limited, which leads to graceful overload behavior.

These factors are further explained below:

Higher Capacity: In TDMA (like GSM) systems, typically a frequency reuse factor of seven is used and the available spectrum has to be partitioned among the seven cells, thereby leading to an inefficient utilization of available channels. In a CDMA system, on the other hand, the entire available frequency spectrum can be used in each cell with increased efficiency. EVRC (enhanced variable rate coding) codec, because of voice activity detection, utilizes smaller bandwidth over the air interface and leads to increased capacity. Furthermore, because the WLL terminals are typically fixed, very accurate power control is feasible, resulting in significant capacity gain.

Greater Range: Range is related to the path loss and the minimum signal level that the receiver can decode reliably. Since in the CDMA network the receiver has the capability to apply a gain factor to the received signal, it can decode weaker signals more successfully, thereby increasing the range.

Sectorization: Because the sectors at a cell site in a CDMA system can use the same frequency, the process of sectorization can increase the system capacity by almost a factor of three (in a three sector antenna), with a modest outlay for additional antenna and RF equipment.

Frequency Planning: As opposed to TDMA and FDMA systems, where extensive frequency planning needs to be deployed for assigning frequencies to various cells, the single-frequency reuse in CDMA systems avoids this step. However, CDMA systems do require allocation (and planning) of long PN codes to individual cells.

Space Diversity: Use of rake receivers in a CDMA system permits the subscriber terminal to simultaneously receive signals from two or more base stations and thus utilize the advantages of space diversity for mitigating effects of multipath fading as well as providing the capability for soft handoffs.

Soft Capacity Limit: The capacity of a cell or a sector in a non-CDMA system is hard-limited in the sense that if all the available traffic channels in the cell/sector are busy, the next attempt is rejected. In a CDMA system, the operator has the flexibility to admit additional users during peak periods by providing a somewhat degraded service (increased bit error rate). This capability is especially important when calls might be dropped during handoff because of a lack of free channels in WLL systems that support mobility.

6.7 RADIO ENGINEERING TASKS FOR WLL SYSTEM DEPLOYMENT

Some of the radio-planning issues were addressed in Chapter 5. From a WLL system planning perspective, the major tasks for radio planning include

- Estimating the required number of cells
- Selection of cell sites and clusters
- Allocation of radio channels to individual cells*
- Establish the frequency reuse plan*
- Establish the antenna configuration

The radio engineering for the deployment of a WLL system is essentially concerned with the planning of the number and location of the cells so as to provide adequate coverage and capacity to meet the traffic intensities and distribution (geographic) needs of the current and forecasted subscriber base. An additional task that the planning engineer needs to perform is to allocate a suitable number of radio channels to the cells and establish frequency reuse plans where applicable (if the number of required cells is small, frequency reuse is not possible). The choice between omnidirectional versus sector antennas (and their number) also needs to be considered.

Figure 6.1 illustrates a simple topography for a WLL system. Though the figure shows circular cells (which will be the case in practice if the base station antenna is omnidirectional) serving a part of the coverage area, one needs to remember that for planning purposes the cells are generally assumed to be hexagonal and nonoverlapping, and it is a common practice to use sector antennas at the base station to increase capacity and coverage for a WLL system.

The total number of cell sites is a critical parameter for the WLL network because it is one of the key cost drivers in determining the total network cost. It needs to be carefully estimated to meet the current and forecasted coverage needs. Although it is possible to add additional cells later, unlike a cellular system, in a WLL system there may be significant cost penalties because adding and/or rearranging cell sites will involve reorienting directional antennas used at the subscriber locations.

In many ways, planning a WLL radio network is similar to planning a mobile cellular network. In each case a cell plan detailing radio coverage and capacity must be produced. In addition, a backhaul, transport, and switching network must be provided to interconnect the base stations. Thus, same or similar (software) radio

*Not required for CDMA-based systems.

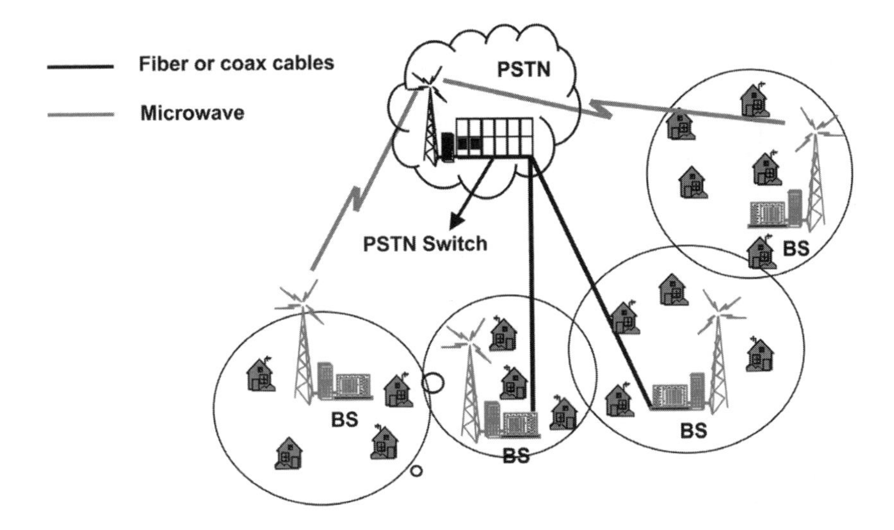

Figure 6.1. A simple topography for a WLL system.

planning/engineering tools utilized for cellular network planning can be used for WLL systems as well. However, there are some basic differences between the coverage and capacity profiles and requirements between WLL and cellular mobile systems.

6.7.1 Coverage and Capacity Profiles for WLL Systems

Generally, the coverage required for a cellular mobile system needs to cover areas where the subscribers are located as well as where they may travel for business and pleasure—even though, as illustrated in Figure 6.2, the subscriber populations may be centered at a few islands. For WLL systems, on the other hand, coverage is required only where the subscribers are located, which may be islands of populations of varying size.

The initial requirement for roll out of a cellular mobile system is usually radio coverage. There is a critical mass in terms of radio coverage at which subscribers realize they can make calls from most of the locations they may find themselves in. Only at this stage does the cellular mobile network become attractive to the potential customers.

On the other hand, in a WLL system (with no mobility) a subscriber will not be concerned about coverage, provided that his own fixed location can be reached. Thus it is possible for WLL operators to gain revenue from a very limited coverage deployment.

In general, the requirement to provide continuous coverage over very large areas may be less prevalent in WLL network design than in cellular mobile network

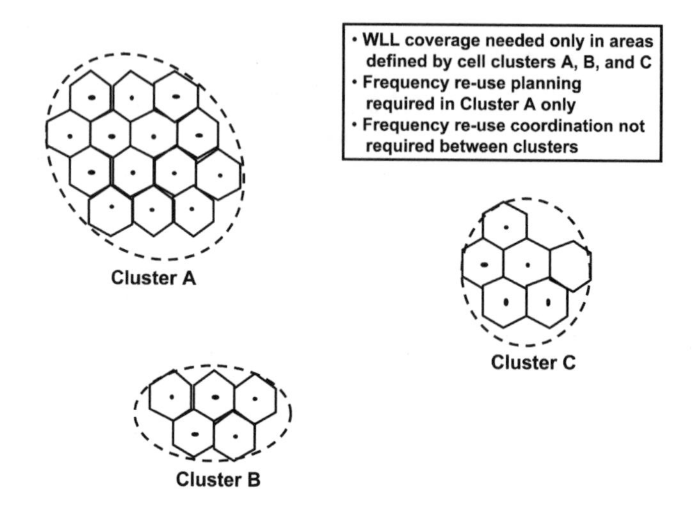

Figure 6.2. Islands of coverage typically encountered in WLL systems.

design. The following two coverage conditions may be encountered in WLL systems:

1. Multicell continuous coverage involving extensive frequency reuse within the coverage area. This condition may be representative of high-capacity networks in urban or metropolitan environments where a large number of small cells are employed.
2. Isolated selective coverage representing coverage in small islands such as key population centers (villages, towns) with areas of limited or no service in between these islands of coverage areas.

As illustrated in Figure 6.2, if the coverage areas are geographically disjointed or they differ significantly in their subscriber densities and/or their usage patterns, the calculations for number of cells required will need to be repeated for each island of coverage area and/or for each distinct subscriber density and usage environment.

6.7.2 Estimating Number of Cells to Meet Capacity Requirements

The total amount of frequency spectrum available to an operator is finite, and this can be divided into a finite number of radio channels based on the WLL technology (radio interface) that the operator has selected. The capacity requirement involves selecting a suitable number of cells (and reusing radio frequency channels as appropriate) so that the system can support the traffic generated by the subscribers in the busy hour while meeting a given service criteria like blocking probability or delay in obtaining access to the network.

For circuit-switched services (e.g., voice, fax) the total subscriber busy hour traffic load can be estimated in erlangs or CCS (hundred call seconds) by multiplying the total number of subscribers and the busy hour erlangs per subscriber. The service criteria or grade of service (GOS) for circuit switched services is generally specified as 1% blocking (i.e., probability that an incoming call to or outgoing call from a subscriber is denied service = 0.01). For data services, appropriate throughput and delay parameters need to be used.

In order to estimate the number of cells required to serve the total traffic, one may assume that each cell carries the same amount of traffic; cell sizes can be suitably varied to meet this assumption. The known parameters include

A = total traffic to be carried by the system (in erlangs)
p = grade of service (e.g., $p = 0.01$, 0.02 blocking)
N = total number of traffic channels available
(which depend on available spectrum and the deployed technology)
m = frequency reuse plan (e.g., $m = 4$, 7, 12)
$j = N/m$ = number of channels in each cell (for cluster size m)

One can now calculate A_c = the traffic carried by j channels in each cell using the Erlang B traffic formula. The number of cells required to carry the total traffic A, which is the same as the number of cells needed to satisfy the system capacity requirements, denoted by N_{cap}, is then given by

$$N_{\text{cap}} = \frac{[\text{Total system traffic } A]}{[\text{Traffic carried by each cell}]}$$

The following simple example illustrates the calculations:

Example (Circuit Switched Services) Consider a WLL system that needs to serve 30,000 subscribers each of which generates an average busy hour traffic load = 0.06 erlangs. The WLL system being deployed is a TDMA system that can provide a maximum of 392 full duplex traffic channels and the system will utilize a frequency reuse factor of 7. The system is to be designed for a 1% blocking grade of service. Thus,

Total system traffic $A = 30{,}000 \times 0.06 = 1800$ erlangs
Number of channels per cell $j = 392/7 = 56$ channels
Grade of service $p = 0.01$

Using the Erlang B traffic tables, the traffic capacity of each cell A_{cl} (with 56 channels and 1% blocking) = 43.3 erlangs, and

$$\text{Number of cells to support capacity requirements, } N_{\text{cap}} = \frac{1800}{43.3} = 42$$

Similar procedures can be used for data services (or a combination of data and voice services) deploying appropriate traffic demand measures (bits/subscriber) and grade of service parameters (throughput, delay).

6.7.3 Estimating Number of Cells to Meet Coverage Requirements

Coverage in a WLL system is related to the range of the system, which in turn is determined by the interference characteristics of the deployed technology and the fade margins applicable. Having estimated the number of cells required to meet the system capacity requirements (see previous section), one needs to estimate the average cell radius r (in km) and cell size or cell area C_a (in km^2), which will depend on such factors as the transmit power, receiver sensitivity, the local terrain, and applicable fade margins.

Having an estimate of the average cell size C_a and the number of cells N_{cap} to meet the capacity requirements, one can verify if their product ($C_a \times N_{\text{cap}}$) equals or exceeds the size of the total coverage area, thereby meeting the system coverage requirements as well. If not, then additional cells are required and the number of cells to meet the coverage requirements (denoted by N_{cov}) can be estimated as

$$N_{\text{cov}} = \text{Number of cells for coverage} = \frac{\text{Total system coverage area in km}^2}{\text{Average cell size } C_a \text{ in km}^2} \times L$$

where L is a factor that reflects the inefficiency introduced by the (unavoidable) overlap between adjacent cells.

In the case of WLL systems the value of L may be higher (closer to 1) than that of cellular mobile systems. For cellular mobile systems, a certain amount of overlap between adjacent cells is desirable in order to ensure proper handover of calls when the traveling subscriber crosses cell boundaries.

6.8 SELECTING CELL SIZES AND CELL SITES

Having estimated the number of cells required, for planning purposes it can be assumed that all cells have the same radius determined by arranging the cells over the system coverage area (i.e., cell size in km^2 = system coverage area in km^2/ number of cells). Placement of cell sites (base stations) is influenced by a number of factors including being able to lease a selected site, local regulations, and restrictions.

If the system is coverage limited (i.e., the number of required cells is determined by the system coverage), then the system is generally designed to provide maximum range. Cell site antennas therefore should optimally be placed on high ground or high buildings. If the system is capacity-limited, the system will use smaller cells with significant frequency reuse. In this case, cell site antennas can be placed on lower buildings (but still avoiding potential obstructions in the line-of-sight paths to the intended subscriber locations) in order to minimize co-channel interference from other cells using the same frequencies.

Planning the placement of cell sites is an iterative process that is generally carried out using some form of a software tool. Initially the coverage is divided into a regular (hexagonal) grid, with the radius of the hexagons based on the calculated radii and base stations located as close as possible to the center of each cell using a typographical map of the coverage area. The initial plan can be refined after site availability, leasing costs, local regulations, and restrictions have been factored in.

As discussed in Chapter 5, sectorization is commonly employed in WLL networks to gain additional capacity and range, especially in an urban or metropolitan environment. Cell site antenna orientation is required for sectorized antennas so that the requirements for coverage and quality of service for all subscribers are met.

6.9 NETWORK INTERCONNECTIONS

The various options available for interconnecting the WLL system base stations to the PSTN were addressed in Chapter 4 (Section 4.5). The available options for transmission medium and protocols for interconnection to the PSTN are summarized below.

Transmission Medium Options

- Coaxial cable
- Fiber links
- Microwave links
- Satellite links

Protocol Options

- Channel associated signaling (CAS)
- Signaling system 7 (SS7)
- Digital subscriber signaling 1 (DSS1 or Q.931)
- ETSI V5.1 interface
- ETSI V5.2 interface

Independent of the transmission medium used to connect the base stations to the PSTN switch, a common protocol needs to be supported at both ends of the link. If the base station and the switch are provided by the same vendor as a package, the vendor may implement a proprietary interconnection protocol. Such proprietary protocols can be more efficient and can save bandwidth over the link. However, the downside is that the operator is locked in to the products provided by the vendor and will have little flexibility in using other vendors' products (even if they are more economical).

In WLL systems where the base stations are to be connected to existing PSTN switches, it is advisable to choose one of the standard protocols listed above. The choice will generally depend on the additional cost for upgrading the PSTN switch and the range of services that the WLL system needs to support.

For example, CAS (channel associated signaling) can only carry PSTN (simple voice) traffic and Q.931 can only carry basic rate ISDN (2B + D). The V5.2 protocol is considered to have a number of advantages in WLL applications. V5.2 protocol, its basic architecture, and its functionality were discussed in Section 4.5. Most manufacturers who have developed WLL systems support V5.2 protocol in their products.

6.10 RANGE AND RELIABILITY ASPECTS OF MICROWAVE LINKS

A microwave point-to-point link is a common option for connecting a base station to the PSTN—especially in WLL systems that provide services to distant rural communities where terrestrial links (coax, fiber) are not viable. Microwave links operate in a number of frequencies with differing ranges. Figure 6.3 illustrates the rapid reduction in the range of a microwave system with the increase in operating frequency of the microwave link.

For distances greater than those achievable by a single link at an available frequency, multihop links can be deployed. This will, of course, lead to increased cost. Typically, the spectrum is crowded at the lower frequency end and it may not always be available. It is generally expected that installing its own microwave links by the WLL operator will not be economically viable (compared to leased lines) for

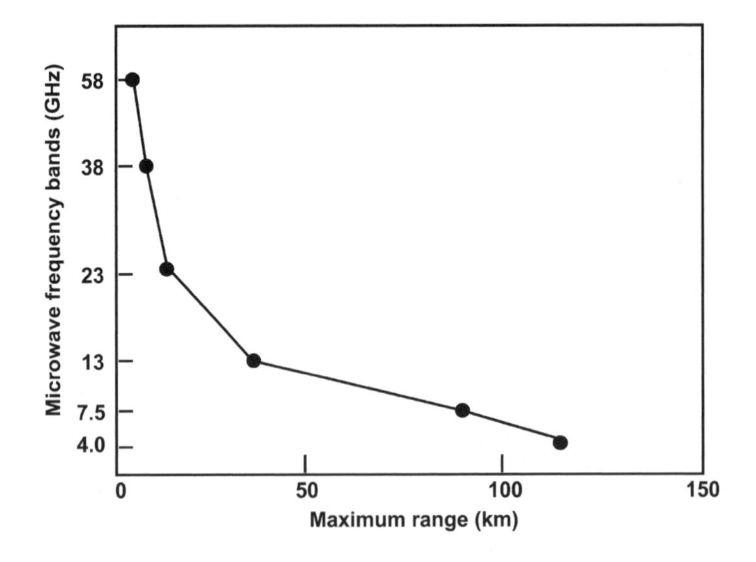

Figure 6.3. Range of microwave links operating at different frequency bands.

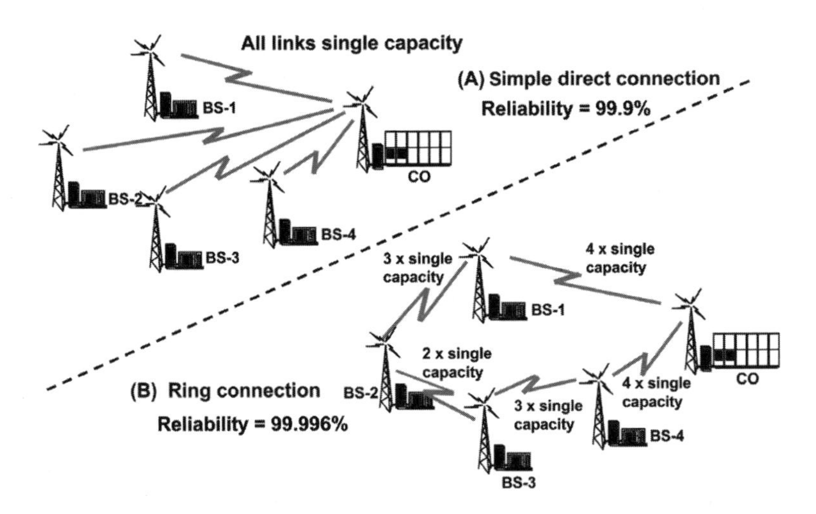

Figure 6.4. Examples of interconnection options for microwave links.

bandwidths below 2 Mb/s or distances below 15 km. For most WLL systems, bandwidths at or above 2 Mb/s are required and microwave links may be an economic alternative.

The overall reliability of the WLL system interconnected using microwave links depends on the interconnection arrangement used. Two alternate arrangements are illustrated in Figure 6.4.

The reliability of the two connection models can be estimated in a simple manner. Thus, in the case of simple direct connections illustrated in Figure 6.4(A), if each link has a reliability of 99.9%, then the base stations in this architectures will also have reliability = 99.9%.

The overall reliability for the ring connection in Figure 6.4(B) can be calculated as 99.996%, illustrating that the ring connection has one of the best reliability performances. However, the gain in reliability for the ring connection is achieved at the cost of additional link capacity that will have to be provided, where the required capacity increases vary from two to four times single link capacity, depending on the closeness of the link to the PSTN switch.

Sometimes in order to achieve very high reliability microwave links, additional radio units need to be placed on the same masts in a *hot standby* mode, thereby almost doubling the equipment cost and thereby significantly increasing the cost.

6.11 BUSINESS ASPECTS AND BUSINESS CASE FOR WLL SYSTEM DEPLOYMENT

An operator who intends to deploy a WLL system needs to verify that the overall enterprise will be economically viable and ultimately profitable. Generally, the vehicle used for this exercise is the preparation of the so-called *business case*.

TABLE 6.5. Example Distribution of Potential Costs for a WLL System

Cost Item	Percent of Total
Interconnect charges	45
Network	22
Staff	11
Remote units	9
Operations	7
Administration	3
Marketing	3

Essentially the business case for a WLL system deployment tries to balance the total potential costs against total expected revenues over a 5- to 10-year planning horizon.

The calculation of total costs and revenues is not very simple, and the preparation of the business case is generally considered an art rather than an exact science. In order to properly weigh various options, in terms of network architecture, technology choice, and associated cost and revenue scenarios, a number of iterations are required before a business plan can be finalized.

Tables 6.5 and 6.6 illustrates (purely as an example) the various cost and revenue items that may be relevant for a WLL project and also lists their relative contributions to total cost or revenue. While some of the items in these tables are self-explanatory, the following items probably need some additional clarification:

Interconnect Charges: In the current environment of deregulation of telecommunication services, most WLL operators are likely to be independent operators. Under this scenario, calls from the WLL system terminating on or transiting through the local PSTN network will be subject to interconnect charges. Conversely, calls terminating on or transiting through the WLL network will derive interconnect revenues.

Remote Units: Remote subscriber units for the WLL system are likely to be provided by the operator on a lease basis, and the operator will incur costs for

TABLE 6.6. Example Distribution of Potential Revenues for a WLL System

Revenue Item	Percent of Total
Call charges	68
Interconnect charges	24
Subscription charges	6
Connection charges	2

purchasing, installing, and maintaining these units. The revenue from the leased units will be reflected in the monthly subscriptions from the customers.

Connection Charges: This is the initial connection charge that covers the labor (and probably) material for initial provision of service at the subscriber premises.

Network Costs: The main components of the network cost include the capital and installation costs associated with

- Base stations
- Base station interconnection (backhaul)
- Base station controllers (if required)
- Base station controller interconnection (if required)
- Switching costs
- Operations, maintenance, and billing system costs

Base Station Costs: These include cost of (a) hardware as quoted by the manufacturer, (b) the installation cost including the cost of acquiring or leasing the site, cost for erecting the mast, and connection to power sources, and (c) planning overhead associated with each base station.

Base Station Interconnection: This is the cost for connecting the base station to the PSTN switch (i.e., the backhaul). The cost will depend on the backhaul technology deployed (i.e., leased line or microwave). For leased lines a one-time connection cost + monthly charges will apply. For microwave there will be full capital cost of the microwave link and its installation cost.

Base Station Controller and Their Interconnection Costs: Some WLL systems deploy BSCs to connect a number of base stations. In this case, the cost of BSCs and their installation and interconnection costs (similar to BS interconnection) should be included.

Switching Costs: Depending on the architecture adopted for the WLL system, the system may or may not deploy a switch. In a smaller system, a BS and/or a BSC may be directly connected to a PSTN switch. In this case the WLL operator will be required to pay switch termination charges (and interconnect charges for intrasystem calls as well). A WLL system serving a large number of subscribers with a significant community of interest may wish to install its own switch. The cost in this case will include hardware costs, the cost of building and its site, the cost of redundant power supplies, and other security costs.

The operation, maintenance, administration, and marketing costs may include a number of items, some of which are indicated below:

- Site rental costs
- Leased line costs
- Maintenance costs

- Radio spectrum costs
- Subscription management costs
- General management costs
- Marketing, sales, and customer retention costs

Site rental costs can vary dramatically according to location (city center versus rural).

Leased line costs are highly dependent on the capacity, quality, length of the lines, and the PSTN operator concerned.

TABLE 6.7. A Sample Business Case Spreadsheet for WLL System Costs

Cost Items (Forecasts)	Year X	Year X + 1	Year X + 2	Year X + 3	Year X + 4
• Total subscribers (K)					
• Total base stations					
• Total base station controllers					
• Total switches					
Network + remote unit costs (capital + installation) in $M					
• Base station costs					
• Base station controller costs					
• Switch costs					
• Remote unit costs					
• BS & BSC interconnection costs					
Recurring costs in $M					
• Radio spectrum leasing					
• Site rental					
• Operations and maintenance					
• Staffing					
• Marketing and sales					
• Building rental + utilities					
• Subscriber billing					
• Total capital + recurring costs in $M					

TABLE 6.8. A Sample Business Case Spreadsheet for WLL System Revenues

Revenue Related Items	Year X	Year X + 1	Year X + 2	Year X + 3	Year X + 4
• Total subscribers (K)					
• Average call minutes per subscriber					
• Net revenue/call minute ($)					
Revenue generated ($M)					
• Total call revenues					
• Total installation + fixed subscription charges					
Total revenues ($M)					

Maintenance costs generally vary with the technology selected for the WLL system. A benchmark figure from the cellular industry that is generally used in WLL systems is in the range of 1.0–2.5% of total capital cost.

Radio spectrum costs are country-dependent. In some countries the spectrum is leased (implying recurring cost), while in others it is auctioned (implying a capital cost).

Subscriber management costs are associated with sending bills to subscribers and managing billing problems.

General management costs include management salaries, cost of HQ building, power/water bills, fleet vehicle costs, and so on. An approximate estimate is around 1% of total revenues.

Marketing, sales, and customer retention costs are an essential component in a competitive environment and are generally estimated as less than 1% of total revenue.

Spreadsheets for the cost and revenue components over the target planning horizon form the basis for preparing the business case. Example spreadsheet formats for total costs and revenues are shown in Tables 6.7 and 6.8, respectively.

The revenue-related spreadsheet illustrated in Table 6.8 applies to only one class of subscribers for whom a particular tariff is applicable (e.g., residential subscribers).

Separate spreadsheets will be required for each class of subscribers that have different tariff structures in terms of call minute charges and monthly subscriptions.

Obviously the above two spreadsheet formats are for illustrative purposes only. In practice, much more detailed and complex spreadsheets are required to prepare a complete business case.

CHAPTER 7

Examples of Commercial WLL Systems

A wide range of WLL systems based on cellular technologies as well as proprietary technologies are available. This chapter provides some representative examples of commercial WLL systems for deployment in various market environments. The reader should note that the descriptions provided here are based on information available in recent publications and literature in the public domain at the time of writing. Since products are frequently updated and replaced by new versions, the reader is advised to contact the manufacturers or their marketing representatives for latest product descriptions and product availability.

7.1 INTRODUCTION

As was mentioned in the preceding chapter, WLL systems can be classified into the following two broad categories:

- Systems that use the same radio transmission technology standards (radio interface) as one of the cellular mobile systems or cordless telecommunication systems
- Systems that use proprietary radio transmission technologies

In the former case, the WLL system is constrained to operate in the frequency band(s) associated with the cellular mobile system or cordless telecommunication system on which the WLL system is based. For example, a GSM-based WLL system is constrained to use 900-MHz, 1800-MHz, or 1900-MHz bands in which the radio interface for current GSM cellular mobile systems has been standardized. For these systems, the operating frequencies are highly regulated, and frequency sharing between a cellular mobile system and a WLL system may be required.

In the case of proprietary systems, there is no restriction on the frequency band for which the system is designed. The vendors may design and develop systems for

Introduction to WLLs. By Raj Pandya
ISBN 0-471-45132-0

specific operators who hold a license for a given frequency band. The frequency bands above 3 GHz are commonly used because they may be easier to acquire from the regulator than cellular mobile bands.

The representative WLL systems described in this chapter include:

No.	System	Manufacturer	Access Type	Technology Type
1	STRAEX	LG Electronics (South Korea)	CDMA/FDD	Cellular standard
2	AirLoop	Lucent Technologies (USA)	CDMA/FDD	Proprietary
3	corDECT	Midas Corporation (India)	TDMA/TDD	Cordless standard
4	Internet FWA	Nortel Networks (Canada)	TDMA/FDD	Proprietary

7.2 STRAEX™ WLL SYSTEM (LG ELECTRONICS, SOUTH KOREA)

The STRAEX WLL system is targeted for an environment where both fixed subscribers and subscribers with limited mobility are to be supported. It is based on existing CDMA radio standards so that it is suitable for rapid deployment and expansion. However, because it is based on existing cellular technology which utilizes low bit rate voice coding technologies, the overall quality (voice quality, bit error rates) will not match those provided by a wireline access network or WLL systems using higher bit rate coding like 32-kb/s ADPCM.

7.2.1 Key System Specifications

As indicated in Table 7.1, this WLL system is available in two distinct frequency bands for which CDMA cellular systems are standardized. The 800-MHz band is

TABLE 7.1. Key STRAEX WLL System Specifications

Parameter	Specification
Access method	CDMA/FDD
Channel bandwidth	1.25 MHz
Frequency bands	800 MHz (TIA IS-95) 1900 MHz (ANSI J-STD 008)
Voice coding	QCELP (8 kb/s & 13 kb/s) EVRC (8 kb/s)
Services	Cellular-type services (including limited mobility)

associated with the IS-95 CDMA radio interface standard specified by TIA in the United States. The CDMA standard for the 1900-MHz band was developed as a joint standard between the TIA and the T1 Committee (ANSI J-STD 008) as part of the PCS specifications.

As discussed in Chapter 3, in a CDMA system the number of subscribers supported by a cell site is variable. Unlike FDMA or TDMA systems, a system based on CDMA technology does not assign a call attempt by any given subscriber a specific portion of either radio frequency or time. All subscriber units transmit and receive at the same time employing different codes, over the same frequencies using the entire 1.25-MHz bandwidth assigned to the carrier.

The number of active calls that can be supported on a radio link (i.e., 1.25-MHz bandwidth) assigned to each cell site (omni-directional antenna) or an individual sector in the cell site (sector antennas) is essentially determined by the target quality of service (voice quality or BER) set by the operator, and it does not represent a hard limit. Typically, a sector is planned to support 56 active calls, and the typical range could be 50+ km under favorable propagation conditions.

7.2.2 System Architecture

The architecture of the STRAEX WLL system shown in Figure 7.1 is very similar to architectures of most WLL systems based on existing cellular technologies. A number of cell sites (base station transceivers) are connected to a base station controller that provides connectivity to public networks (PSTN, PSPDN/Internet).

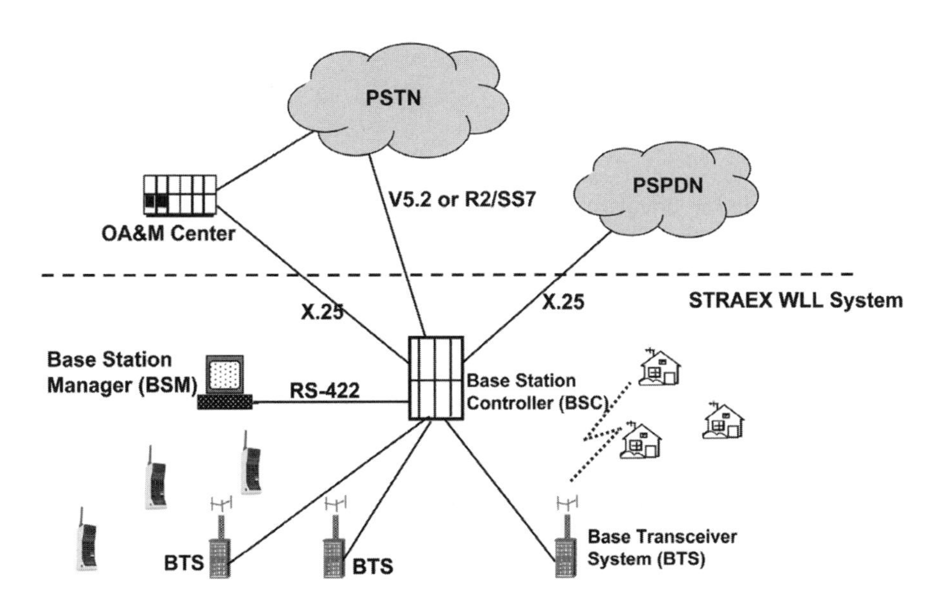

Figure 7.1. STRAEX WLL system architecture.

This system provides two options for interconnection to the PSTN. Either a V5.2 interface or a R2/SS7 signaling interface is available. The configurations and capacities are affected by the choice of the interface. For the V5.2 option, the system can support a maximum of 32 E1 links to the PSTN (as opposed to 64 E1 links for the SS7 option).

7.2.3 Service Features and Capabilities

The broad service features provided in the STRAEX WLL system include:

Voice: The voice coding methods available include QCELP (8K and 13K) and EVRC (enhanced variable rate coding) so that the system can provide cellular-grade voice quality.

Custom Calling Services: For the V5.2 interface type system, all custom calling services provided by the PSTN can be provided transparently to the WLL subscribers. For the R2/SS7 interface option, these services are provided from within the WLL system by enhancing the functionality of the BSC to a local switch level. The type of custom calling services supported by STRAEX WLL system include:

- Call forwarding
- Call waiting
- Call transfer
- Caller line identification presentation
- Caller line identification restriction
- Three-way calling/conference call
- Malicious call trace
- Caller authentication

Data Services: Circuit-switched data service at 14.4 kb/s (maximum) is supported. The circuit-switched data services architecture for the STRAEX WLL system is illustrated in Figure 7.2. The BSC supports a pool of modems and interworking functions (IWFs) for data rate adaption, so that circuit-switched services can be provided to both fixed and mobile subscribers supported by the WLL system. It should be noted that the short message service (SMS) is not supported by the STRAEX WLL system. Support of SMS will require additional functionality and equipment in the network (i.e., functionality to store and forward messages).

Mobility: As an option, the system can provide limited mobility, which allows the subscribers to move within the area served by a single BSC. Thus the mobile subscribers in the WLL system are allowed to move between cells (served by the same BSC) and make and receive calls. For incoming calls the paging is universal (i.e., the BSC requests all cells to broadcast the paging signal) and the concerned subscriber will respond to the page. For

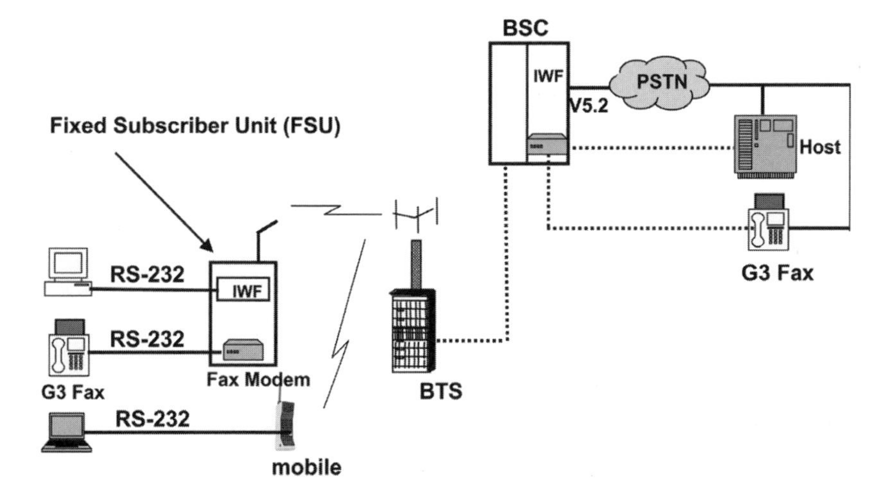

Figure 7.2. Circuit-switched data services architecture for the STRAEX WLL system.

outgoing calls, the BSC will authenticate the subscriber by checking the mobile subscriber unit's ESN (electronic serial number). If the BSC is satisfied that the ESN is not fraudulent and that the subscriber is entitled to mobile service, it will initiate the necessary process for call establishment.

If a subscriber with a call in progress moves from one cell to another or from one sector (in a cell) to another, the system will initiate a handoff process. In a CDMA system, intercell handoffs are soft handoffs (new connection is made before the old connection breaks). However, intersector handoffs are not soft handoffs. Since the system reserves a certain number of channels in each cell and each sector for call handoffs, the capacity of a WLL system which provides limited mobility is reduced (relative to a system without mobility). The intercell and intersector handoffs are illustrated in Figure 7.3.

7.2.4 BSC and BSM Functions and Capabilities

The STRAEX WLL system provides the choice of two types of base station controllers. BSC Type 1 supports the V5.2 interface with a maximum of 32 E1 lines for interconnection to the PSTN and has the following characteristics:

- It acts as an access network (no local switching).
- It depends on PSTN switch for services and call switching.
- It supports 10,000 subscribers (at 0.1 erlang/subscriber and 1% blocking).
- It supports 256 data subscribers (as part of above).
- It has maximum capacity of 995 erlangs and 9950 subscribers.

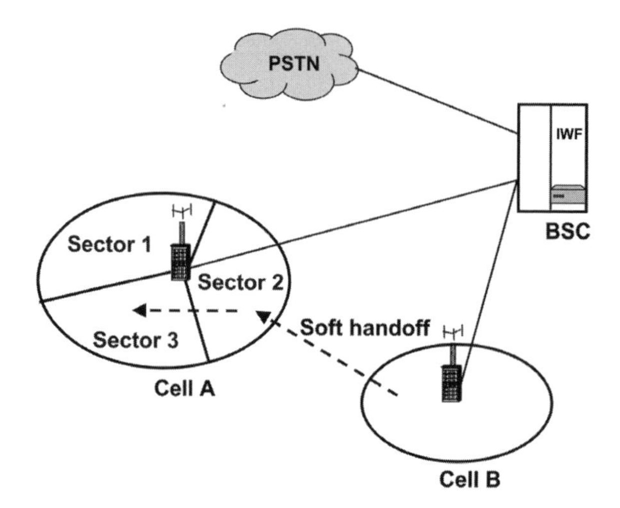

Figure 7.3. Intercell and intersector handoffs in the STRAEX WLL system.

BSC Type 2 supports R2 or SS7 interface with a maximum of 64 E1 lines for interconnection to the PSTN and has the following characteristics:

- It provides local switching capability (acts as a local exchange) and services support for WLL subscribers.
- It supports up to 1920 voice channels.
- It supports 18,920 subscribers (at 0.1 erlang/subscriber and 1% blocking).
- It supports 512 data subscribers (as part of above).
- It provides local accounting and billing functions.
- It has a maximum capacity of 1892 erlangs and 130,000 BHCA.

The broad functions and capabilities supported by the BSC in the STRAEX WLL system include the following:

- Provides interface to the PSTN (V5.2 or SS7 or R2 signaling interface)
- Controls one or more BTSs (max 48)
- Call processing between the WLL network and the PSTN
- Manages intercell handoffs where limited mobility is provided
- Provides trans-coding between 64-kb/s PCM data and 8- to 13-kb/s vocoder data
- Echo canceling
- Subscriber management (authentication, paging)
- Custom calling services (for SS7 or R2 interface only)
- Interworking and modem functions for data services
- OA&M support through BSM (base station manager)

7.2.5 BTS Functions and Capabilities

The STRAEX WLL system provides for three different types of BTS. The compact indoor-type BTS, which is a standard offering, supports up to eight frequency assignments (FAs), each with a bandwidth of 1.25 MHz, and supports three sectors/FA. The outdoor-type BTS supports up to four FAs and supports three sectors/FA, and the third type of BTS (micro/pico-cell BTS) supports up to three FAs and omnidirectional antennas only. Each type of BTS provides the following functions and capabilities:

- Provides radio interface between BSC and subscriber units (TIA IS-95A CDMA standard for 800-MHz band or ANSI J-STD-008 CDMA standard for 1900-MHz band)
- Call control and call resource management
- Digital signal processing for the radio interface
- RF functions like modulation/demodulation, source and channel coding, etc.
- Supports up to 40 channels/sector for 8-kb/s vocoder (23 channels/sector for 13-kb/s vocoder
- Supports up to 29 erlangs of traffic and 870 subscribers at 1% blocking

7.2.6 Remote Subscriber Units

The STRAEX WLL system offers various types of remote subscriber units. Besides the three subscriber units illustrated in Figure 7.4, a multisubscriber unit is also

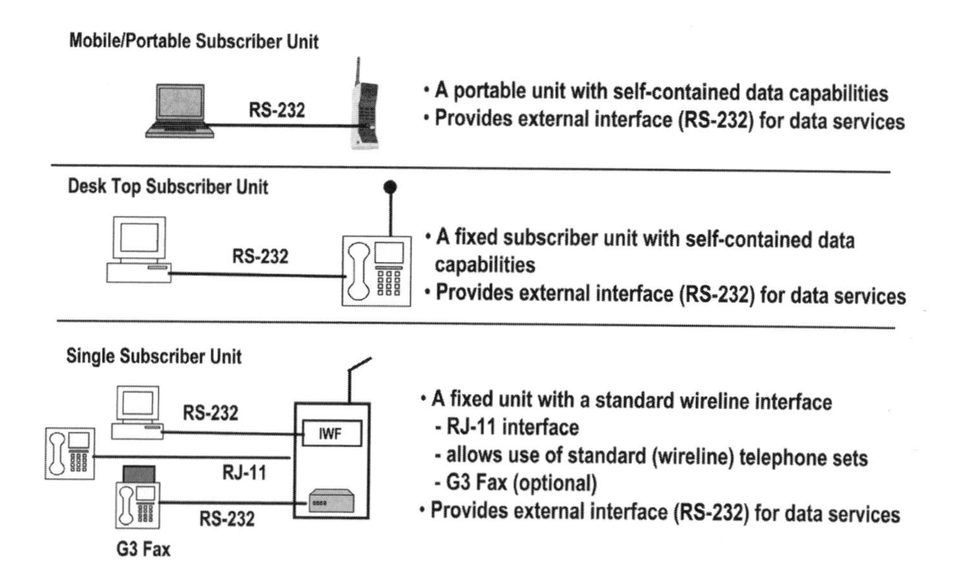

Figure 7.4. Different remote subscriber unit types for the STRAEX WLL system.

available which is a fixed unit with multiple subscriber interfaces. It provides a maximum of 32 subscriber interfaces using four RF channels.

7.2.7 Base Station Manager (BSM) and OA&M Functions

The base station manager (BSM), which provides system management support for one or more base station controllers, provides a range of management functions that include

- User interface to the BSC
- Operations, administration, and maintenance
- System configuration
- S/W downloading
- Supervision of system status
- Interface to the main OA&M Center through TCP/IP over X.25

The operation, administration, and maintenance functions performed by the BSM include:

Initialization:	Downloading of S/W to the system
Fault management:	Detection, isolation and recovery of faults
Configuration management:	Management of data for the system configuration
Security:	Prevention of illegal access to the system
Performance management:	Collection of data on system utilization and performance
Accounting and billing (SS7 and R2 I/F only):	Collection of call detail records and transfer of data to the billing center or the OMC

7.3 AIRLOOP™ WLL SYSTEM (LUCENT TECHNOLOGIES, USA)

The AirLoop System is a proprietary WLL system that utilizes direct sequence code division multiple access (DS-CDMA) technology in the 3.4-GHz frequency band. It claims to offer increased subscriber capacity along with the same level of quality, service, and functionality as enjoyed by subscribers connected directly to wireline systems. The system offers basic analog services, such as telephone, fax, and data transmission, with voice coding rates from 16 to 64 kbps as well as ISDN basic rate access (BRA) services providing up to 128-kbps data circuits. Packet data are also supported which provide direct connection (without a modem) to the internet.

7.3.1 AirLoop WLL System Radio Specification

The AirLoop system operates in the 3.4- to 3.6-GHz frequency band with 100-MHz duplex spacing. The 3.4-GHz band plan supports 10 RF channels, with each RF channel requiring a pair of 5-MHz links (one for the up-link and one for the down-link). A system operating with 90° sector antennas and a frequency reuse of two requires two channels (i.e., an RF bandwidth of 10 MHz in each direction). As subscriber densities increase, the option of using additional RF spectrum instead of additional cells is an option in the AirLoop system. The key radio specifications for the AirLoop system are captured in Table 7.2.

7.3.2 System Architecture for the AirLoop WLL System

The architecture of the AirLoop WLL system shown in Figure 7.5 is very similar to architectures of most WLL systems based on existing cellular technologies. However, the AirLoop system uses proprietary terminology. For example, CATU and CRTU perform functions similar to those of a base of station controller (BSC) and a base transceiver station (BTS), respectively. A number of cell sites (CRTUs) are served by a CATU that provides connectivity to public networks (PSTN, PSPDN/Internet). Either V5.1/V5.2 interface or CAS signaling interface can be used for interconnection to the PSTN.

The major elements of the AirLoop architecture consist of

- Central access and transcoding unit (CATU), located in close proximity of the local exchange and supports V5.1/V5.2 or CAS interface to the LE
- Central transceiver unit (CTRU) located at the base station site and connected to the CATU using E1 links
- Subscriber's transceiver unit (STRU), located on the outside of the subscriber premises

TABLE 7.2. Key Radio Specifications for the AirLoop WLL System

Parameter	Specification
Access method	DS-CDMA
Duplexing method	FDD
Operating frequencies—A band	
Subscriber-to-base (GHz)	3.40–3.45
Base-to-subscriber (GHz)	3.50–3.55
Operating frequencies—B band	
Subscriber-to-base (GHz)	3.45–3.50
Base-to-subscriber (GHz)	3.55–3.60
Duplex separation (MHz)	100
RF channel bandwidth (MHz)	5
Total number of RF duplex channels	115
Modulation method	QPSK (spreading)
Speech coding methods	LD-CELP (16 kb/s) ADPCM (32 kb/s) PCM (64-kb/s mu law)

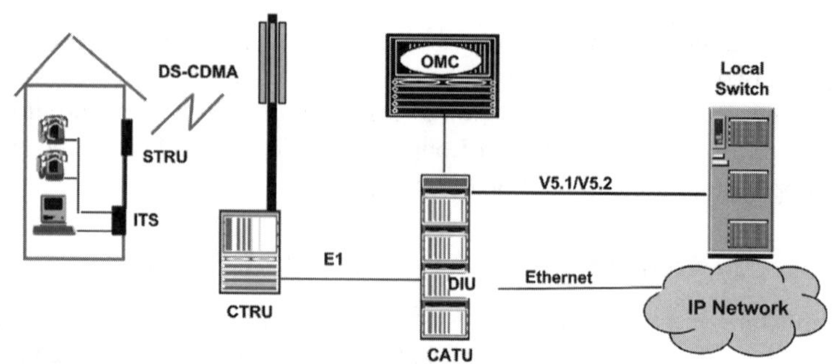

CATU: Central Access & Transcoding Unit (BSC)
CTRU: Central Transceiver Unit (BTS)
DIU: Data Interface Unit
OMC: Operations & Maintenance Center
STRU: Subscriber Transceiver Unit (Outdoor Unit)
ITS: Intelligent Telephone Socket (Indoor Unit)

Figure 7.5. System architecture for the AirLoop WLL system.

- Intelligent telephone socket (ITS), generally located inside the subscriber premises
- Operations and maintenance center (OMC), which manages all aspects of the system's operation using LAN connections to appropriate network elements

The functions provided by these components are further described in the next section.

7.3.3 AirLoop System Components and Their Functionalities

Central access and transcoding unit (CATU) provides the link between the local exchange and the transceiver units. The CATU communicates with the local exchange with up to 16 2.048-Mb/s interfaces that support V5.1 and V5.2 protocols. Support for channel associated signaling (CAS) is also available for interconnection to nondigital local exchange types. The functions supported by the CATU include:

- Transcoding from 64-kb/s A-law coded speech to 32-kb/s ADPCM or 16-kb/s LD-CELP
- Controlling the allocation of radio resources
- Providing the operations, administration, and maintenance support
- Controlling and maintaining of subscriber access

Data interface unit (DIU) provides the link between the CATU and the IP network access point. The DIU provides the interface for packet data subscribers between the AirLoop network and the Internet service provider (ISP). Multiple subscribers are connected to an ISP using Layer 2 tunneling protocol (L2TP). The DIU routes all packet data calls to the IP network, diverting them away from the local exchange (LE).

Central transceiver unit (CTRU) transfers analog and digital subscriber signaling between the radio interface and the CATU. The CTRU is connected to the CATU via a concentrated 2048-kbps E1 link using a proprietary protocol to transfer both signaling and user data. The CTRU also provides access for a local maintenance terminal (LMT) and local wireline connections. RF coverage is generally provided with 90° sector antennas with a frequency reuse of 2.

Subscriber's transceiver unit (STRU) is the outdoor unit that houses a radio transceiver, a directional antenna, a modem, control circuitry, and a connection to the ITS. The directional antenna provides increased system range and resistance to fading and interference. The STRU is mounted on an outside wall or on a pole on the roof of the subscriber premises. It transfers analog and digital subscriber signaling between the air interface and the ITS and provides timing and alarm information to the network controller.

Intelligent telephone socket (ITS) provides the physical interface to the customer premises equipment. Several telephone socket options are available which are powered by AC mains and provide DC power to the STRU. Backup battery power is also provided. The following versions are available:

- One-line and two-line POTS (indoor unit)
- Two POTS or ISDN (indoor unit)
- Packet data providing an RS232 serial data port and a voice telephone port (indoor unit)
- Multiline (outdoor unit supporting 2, 4, or 8 lines)

The multiline ITS can provide different mixes of voice and data lines (with appropriate coding options), with examples being

- POTS lines with 16-kb/s vocoder
- Four POTS lines with 32-kb/s ADPCM
- Two POTS lines with 64-kb/s PCM
- Four POTS lines with 16-kb/s vocoder + one data line with 64-kb/s PCM for V.90 modem data

Operations and maintenance center (OMC) provides the control and monitoring of the entire system from a single OMC. The OMC provides service provisioning, network management, and alarm monitoring. It is available in varying capacities to support systems from under 50 to over 250 cell sites. The OMC uses SNMP protocol running over a LAN and has a graphical user interface (GUI). The OMC capabilities include

- Service provisioning
- Configuration control
- Software configuration management

- Alarm monitoring
- Security management
- Performance monitoring
- Monitoring of all 2048-kb/s links and subscriber ports

7.3.4 AirLoop WLL System: Services and Capabilities

The AirLoop system is designed to deliver transparent access to services offered by the local exchange switch using the V5.1/V5.2 interface. The range of services that are available to the AirLoop WLL subscribers are summarized below.

Analog Subscriber Services

- Speech (with PCM, ADPCM, or LD-CELP coding)
- Emergency calls priority
- Dual-tone multifrequency (DTMF) signaling
- Voice band data including V.90 transparently (requires PCM speech coding; lower rates are available with ADPCM)
- Group 3 fax, up to 14.4 kbps
- All supplementary services supported by hook-flash and DTMF tones
- 12-kHz or 16-kHz subscriber pulse metering (SPM)

Digital Subscriber Services

- 3.1-kHz audio
- Circuit mode speech
- 64-kbps unrestricted digital access
- Emergency calls priority

Supplementary Services Supported by the Switch

- Calling line id (presentation and restriction)
- Call forwarding/waiting/holding
- Malicious call id
- Direct dialing in (DDI)
- Multiple subscriber numbering (MSN)

Packet Data. The AirLoop packet data feature provides access to IP services at speeds of up to 128 kb/s. This feature allows the user to make voice calls and browse the internet simultaneously. It eliminates the need for a modem or other terminal adapter.

Emergency Calls. The AirLoop system provides priority access to emergency calls. Nonemergency calls may be dropped to connect an emergency call. Up to four emergency numbers can be specified and are software configurable.

TABLE 7.3. Nominal AirLoop System Capacity for a Single Network Equipment Unit

Coding	Bandwidth	Number of Circuits	Traffic Capacity (1% Blocking)
LD-CELP	16 kb/s	115	982 erlangs
ADPCM	32 kb/s	57	442 erlangs
PCM	64 kb/s	28	186 erlangs
B + D	64 + 16 kb/s	23	145 erlangs
2B + D	128 + 16 kb/s	12	59 erlangs

7.3.5 AirLoop WLL System Capacity

The AirLoop system capacity is determined by the radio interface and by the interface to the local exchange. The radio interface has a maximum capacity of 2 Mb/s, split into 128 channels of 16 kb/s, of which 115 are available for subscriber traffic. These 16-kb/s channels can be aggregated to provide data bandwidths as required by the service provided to the subscriber (i.e., 32 kb/s, 64 kb/s, or more for ISDN services, i.e., 64 + 16 kb/s for B + D or 128 + 16 kb/s for 2B + D).

The AirLoop system is highly scalable, and the system capacity can be increased in steps by adding appropriate network elements. The nominal capacity for a single network equipment unit (one CATU and one CRTU) is illustrated in Table 7.3.

The packet data feature of AirLoop system allows data calls with long holding times (such as Internet access) to use only the RF resource required to transfer the data requested. This leaves the normally idle resource free to support other subscriber services. This feature will allow the future growth of packet data or Internet traffic to be handled more efficiently than with traditional circuit-switched techniques using modems. A subscriber terminal capable of supporting one voice channel and a direct data connection using RS232 to a PC is supported.

The coverage provided by a single base station in the AirLoop system varies with the subscriber environment. The coverage for a single cell site can vary from a radius of about 3.4 km for urban applications to 8.9 km for rural applications.

7.4 corDECT WLL SYSTEM (MIDAS CORPORATION, INDIA)

7.4.1 Radio Interface Characteristics for corDECT WLL System

The corDECT WLL system is based on the digital enhanced cordless telecommunications (DECT) standard developed by ETSI and described in Chapter 3 (Section 3.7). The DECT standard defines the radio interface between a fixed part (FP) and a portable part (PP) using a TDMA/TDD access method. The basic properties of the radio interface for the corDECT WLL system are summarized in Table 7.4.

TABLE 7.4. Radio Interface Characteristics for corDECT WLL System

Parameter	Value
Multiple access method	TDMA
Duplexing method	TDD
Transmit/receive frequency band	1880–1900 MHz
Duplex separation	Not applicable
RF carrier spacing	1728 MHz
Number of carriers	10
Number of TDD channels/carrier	12 (full duplex)
Modulation method	GFSK
Speech coding method	ADPCM (32 kb/s)

As indicated in the above table, the 20-MHz frequency spectrum available in the system is used to define 10 RF carriers. Each carrier is then divided into 24 time slots (a pair of time slots defining a full duplex channel), thus providing a capability of 120 full duplex channels. The TDD channel structure for the DECT system was described in Chapter 2 and is illustrated in Figure 2.5. The detailed architecture and functions for the DECT system were described in Chapter 3 and are illustrated in Figure 3.17.

In the DECT system no specific channel is assigned to any DECT terminal-to-base station communication. The terminal can select any of the 120 channels to the base station using the dynamic channel selection (DCS) algorithm. The algorithm requires the terminal to continuously measure the received signal strength indicator (RSSI) on all the 120 channels and then locks on to the base station with the strongest signal. The base station maintains a table of RSSI for all other channels. The table is periodically updated; and if a channel with better RSSI becomes available during call in progress, the call is handed over to the channel with a stronger signal.

7.4.2 corDECT WLL System Architecture

As illustrated in Figure 7.6, the system architecture for the corDECT system consists of the DECT interface unit (DIU) which supports a number of *compact* base stations (CBSs). A CBS serves either fixed or mobile (limited mobility) subscribers. The system also supports relay base stations (RBSs) and a base station distributor (BSD) which can connect up to four remotely located compact base stations to the DIU.

Whereas a relay base station is used to increase the coverage area (up to a radius of 25 km) of an individual CBS, a BSD is deployed to increase the distance between the DIU and a CBS. The BSD can be connected to the DIU through a fiber or a microwave (radio) link. Whereas the CBS connected to a DIU are powered by the DIU, the BSD has to be locally powered. The CBS served through a BSD are powered by the latter.

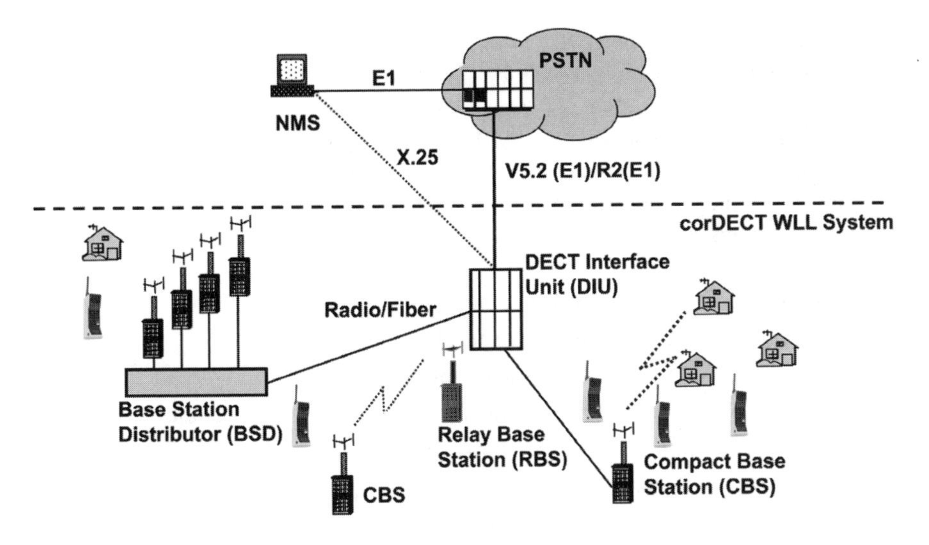

Figure 7.6. corDECT WLL system architecture.

7.4.3 Functions and Capabilities of corDECT WLL Network Components

A *DECT interface unit (DIU)* acts as a switch or remote switching unit (RSU) and interconnects to the PSTN using 2.048-Mb/s E1 lines using V5.2 or R2 signaling interface. Compact base stations are connected to the DIU using three pairs of twisted copper wires. A single DIU can support up to 1000 subscribers at 1% blocking with four E1 links to the PSTN. Multiple DIUs can be connected through a transit switch to further increase system capacity. The DIU can be deployed in one of the following configurations:

- As a local switching unit with R2 signaling connection to PSTN using E1 links
- As an RSU with V5.2 signaling connection to PSTN on E1 links
- As an in-dialing PBX connected to an exchange using two-wire junction lines
- Connected to PSTN using R2 signaling over E1 links

The DIU supports the following functions:

- Interfacing with the PSTN
- Call processing and mobility management
- Supplying power to CBSs
- Transcoding between ADPCM and PCM formats
- Handling DECT network and link layer functions

A *compact base station* (*CBS*) provides wireless access to its serving area. It is powered by the DIU or the base station distributor, depending on the configuration. Each CBS can support about 50 subscribers generating a nominal traffic of 0.1 erlangs with a grade of service of 1%. The coverage radius provided by a CBS can vary from a few hundred meters to 5000 m (4 km), depending on the application (urban versus rural). The main features and operating characteristics for the CBS are summarized below:

- Small pole-mounted or wall-mounted unit serving one cell site
- Supports up to 12 simultaneous speech channels
- Supports up to 50 subscribers at 1% blocking
- Cell radius is between 1.5 and 4.0 km (depending on propagation conditions)
- Deploys two antennas for receiver space diversity
- Directional antennas or omnidirectional antennas can be used
- Omnidirectional antennas with 2-dB, 4-dB, or 6-dB gains are available
- Connection to the DIU uses three standard subscriber pairs from an existing loop plant
- The three pairs carry four ADPCM channels each + signaling + power to CBS (from DIU)
- Max distance from DIU = 4 km (extendable by using a BSD)
- Max distance between BSD and CBS = 0.6 km (0.6-mm twisted-pair copper cable)
- Implements DECT modem in software
- Implements the DECT physical, MAC, and (some) DLC functions
- Software can be upgraded from the DIU
- Performance can be monitored from the NMS

A *base station distributor* (*BSD*) is an optional network element that can be deployed when a cluster of compact base stations are to be located at some distance from the DECT interface unit. It supports up to four remotely connected compact base stations using an E1 link on microwave radio or fiber. It also provides power to the CBSs it serves.

A *remote base station* (*RBS*) is a relay device that is used to extend the coverage range of a compact base station to 25 km.

A *network management system* (*NMS*) manages multiple DIUs and their associated CBSs, BSDs, and RBSs as well as the remote subscriber terminals supported by corDECT system. The functions provided by the NMS include the following:

- It manages up to 30 DIUs.
- It interfaces to the PSTN through an E1 link.
- It can be located anywhere in the network.
- A 64-kb/s X.25 link is available between the NMS and a DIU it manages.

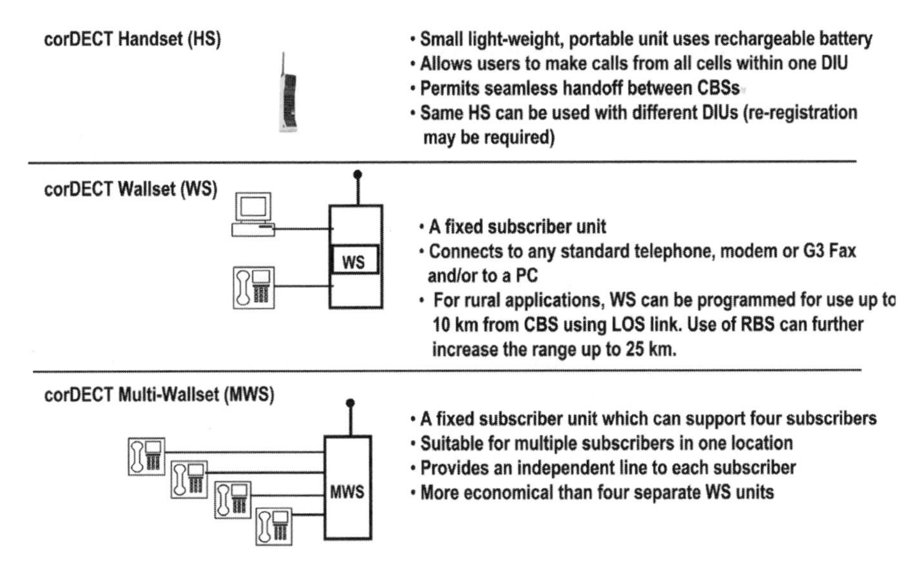

Figure 7.7. corDECT remote subscriber unit options.

- It performs O&M functions for all DIUs and their subscribers.
- It monitors performance of all DIUs, CBS, and subscriber units.
- It provides billing function (when DIUs are configured as independent exchanges).

7.4.4 corDECT Remote Subscriber Unit Options

The corDECT WLL system offers a number of options for subscriber units such as a handset (HS), which is a portable unit that allows the subscriber to make calls from within the coverage area of any of the CBSs connected to the same DIU, a wallset (WS), which is a small wall-mounted unit with an outdoor antenna powered by A/C mains (with a battery backup), and a multi-wallset (MWS), which can cater to up to four independent subscriber lines. The various options and their features are summarized in Figure 7.7.

7.5 INTERNET FWA™ WLL SYSTEM (NORTEL NETWORKS, CANADA)

Nortel Networks' Internet fixed wireless access system is a fully digital wireless local loop system, and it operates in the 3.4- to 3.6-GHz frequency band using TDMA/FDD radio interface. It is designed to provide unlimited *always-on* Internet access, unlimited *always-on* e-mail access, and data rates up to 192 kb/s. Each subscriber can be offered one or two voice lines plus two high-speed data Internet access connections with voice quality equivalent to fixed network connections.

7.5.1 Internet FWA Radio Interface Characteristics

The radio interface characteristics for the Internet FWA system are summarized in Table 7.5. Using this radio interface, the Internet FWA system can support peak data rates of 96 kb/s or 192 kb/s and an analog line interface with 32-kb/s ADPCM voice or 64-kb/s PCM for fax or voice band data.

7.5.2 Internet FWA System Architecture

The basic architecture of the Internet FWA system is illustrated in Figure 7.8, which consists of a network of base stations that are either connected to the PSTN through a switch or connected to the Internet via a router. The base stations communicate with the remote subscriber equipment over the radio interface. The network management center (NMC) provides the OA&M support to the network elements.

7.5.3 Internet FWA System Components

Base station equipment consists of radio transceivers, control electronics, power supplies (including back-up), and an interface to the PSTN switch and the antenna. The base stations can be engineered to serve from sparsely populated rural areas to urban subscriber environments with a coverage range from 400 m to 40 km. It has a modular and scalable design so that an initial system installed to serve low traffic density an be expanded for greater traffic handling capacity as demand grows. It can thus be scaled from an initial design to serve 320 lines to a design that can serve up to 2000 lines. The base stations are connected to the PSTN using standard G.703/G.704 links using either fiber, twisted pair copper wire, or microwave facilities. Options available for signaling interface include V5.2, CAS, 2-wire, and a Nortel network proprietary interface (DMSX).

The *remote service system* (*RSS*) is the equipment located at the customer's premises, and it can provide connections to two standard telephones and data/fax

TABLE 7.5. Internet FWA Radio Interface Characteristics

Parameter	Value
Multiple Access method	TDMA
Duplexing method	FDD + TDD
Duplex separation	100 MHz (also 50 MHz)
Operating frequency band	3.4–3.6 GHz
Down-link	3.402–3.475 GHz
Up-link	3.502–3.575 GHz
RF channel spacing	307.2 kHz
Total number of RF channels	54
Modulation method	pi/4-DQPSK
Speech coding method	32-kb/s ADPCM

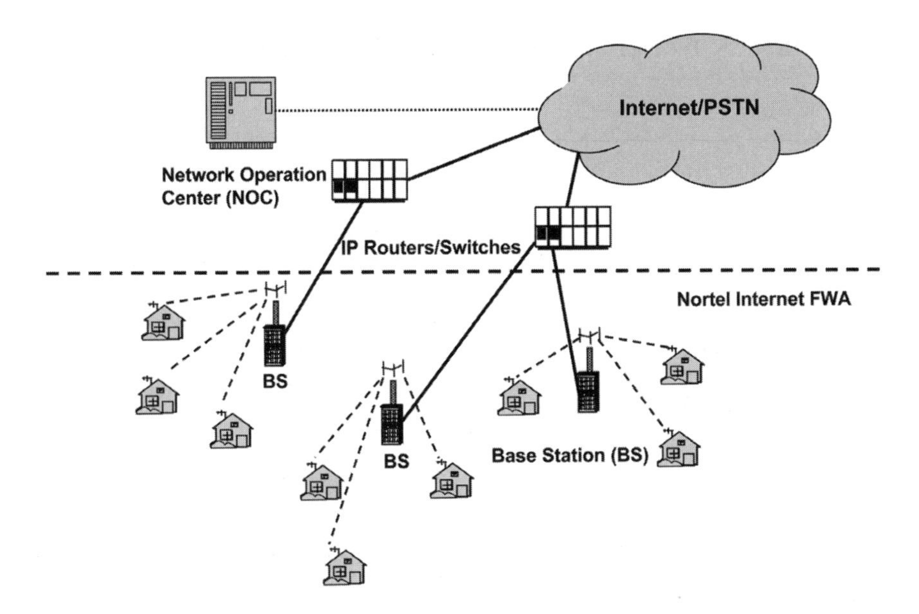

Figure 7.8. Internet FWA system architecture.

equipment. The RSS has the following two components representing an outdoor and an indoor unit which are connected through a remote distribution cable (RDC):

- A *remote transceiver unit* (*RTU*) is the outdoor unit which includes the antenna. It contains the radio transceiver, the processor, and the line interfaces that are mounted behind an octagonal directional antenna. The unit measures around 30 cm in diameter and can be wall-mounted to provide a near line-of-sight path to the base station. It is powered from the indoor unit over the remote distribution cable.
- A *remote power and connection unit* (*RPCU*) is the indoor unit that provides the power and appropriate line terminations. The unit provides 48-V DC power using either 110-V or 220-V AC mains supply with a battery back-up. Solar power equipment is available as an option.

The RSS can provide up to three 32-kb/s time slots for carrying user traffic. A voice call utilizes one time slot using ADPCM coding. In order to accommodate a voice call (32 kb/s) and a voice-band data call (64 kb/s), all three time slots will be utilized. For ISDN service (2B + D), two RTUs will be required.

7.5.4 Internet FWA System Capabilities and Features

The system capacity and coverage range for the Internet FWA system are summarized in Table 7.6.

TABLE 7.6. Internet FWA System Capacity and Coverage Range

Parameter	Value
Service data rates	
Internet interface	96 or 192 kb/s (peak)
Analog line interface	
Voice calls	32-kb/s ADPCM
Fax or voice-band data	64-kb/s PCM
Cell-site capacity[a]	
Subscribers	2000+
Traffic (PSTN only)	220 erlangs
Traffic (PSTN + packet data)	70 erlangs
Lines per user equipment	
Residential lines	2 (max)
Business lines	4 (max)
Party lines	8 (max)
Simultaneous voice calls (PSTN)	256
Nominal coverage (line-of-sight)	
Urban	0.2–2.0 km
Suburban	2.0–8.0 km
Rural	8.0–20 km
With extended range S/W	Up to 40 km

[a]Depends on configuration and grade of service.

Additional features and capabilities of the Internet FWA system include the following:

System Interfaces Supported

- Data networking based on IETF protocol standard over E1 links
- Interconnection to PSTN exchange using V5.2, CAS, 2-wire, or DMSX (proprietary)
- Radio interface (base station–remote user) using TDMA/FDD
- RJ-11 line interface at customer premises
- RS-232 data port connection

Packet Data Services

- Provided through RS-232 data port
- Point-to-point connection compatible with PPP stacks in most PCs
- Runs Internet and Intranet applications transparently
- Supports TCP/IP/PPP, L2TP, DNS Internet standards
- Supports IP services including virtual private networking (VPN)

TABLE 7.7. Key Characteristics of Some Additional WLL Systems

WLL System	Access Method	Frequency (MHz)	Voice Coding	Services supported	Coverage (max)
Proximity I/II (Nortel)	TDMA/FDD	3400–3600	ADPCM	Voice, data, fax, ISDN	15 km
SR500 (SK Telecom)	TDMA/TDD	340–380, 1850–1990	ADPCM	Voice, data, fax	10 km
SWING (Lucent)	TDMA/TDD (DECT)	1880–1930	ADPCM	Voice, data, fax, ISDN	15 km
QCTel (Qualcomm)[a]	CDMA/FDD	800–900, 1800–2000	8-kb/s vocoder	Voice, data, fax	25 km
W-CDMA (LGE)[a]	Wideband CDMA/FDD	2300–2400	ADPCM	Voice, pkt. data, fax	30 km
Airspan (DSC)	CDMA/FDD	2000-MHZ band	ADPCM	Voice, data, fax, ISDN	25 km
NEAXSTAR (NEC)	CDMA/FDD	800- or 1900-MHz bands	8-kb/s vocoder	Voice, 64-kb/s pkt data, G3 fax	25 km

[a]Can support limited mobility.

Other Service Features

- Supports pulse or tone dialing
- Provides wireline voice quality
- Provides dial tone from the local exchange
- Compatible with V.90 56-kb/s modem
- Supports Group 3 fax
- Supports payphones
- Supports PBX, Centrex, and CLASS features

7.6 SUMMARY OF PARAMETERS FOR SOME ADDITIONAL COMMERCIAL WLL SYSTEMS

Besides the four examples of commercial WLL systems described in the preceding sections, there a number of additional WLL systems that have recently been in the market. These utilize a range of technologies and provide differing coverage and capacity. Key characteristics of some of these WLL systems are summarized in Table 7.7.

CHAPTER 8

Broadband Wireless Access (BWA) Systems

Broadband wireless access (BWA) has emerged as an economically viable alternative to such wire-based broadband access technologies as fiber, xDSL, and cable modems for serving residential, small- to-medium-size enterprises (SME) and small office/ home office (SOHO) markets. They are being offered as suitable technologies to support applications like interactive video distribution, high-speed Internet access, and multimedia services. Besides the currently deployed multipoint distribution systems like LMDS (local multipoint distribution system) and MMDS (multichannel multipoint distribution system), there are two other technologies for BWA systems that are being developed. The first of these originated from existing standards on wireless local area networks (namely, IEEE 802.11 and ETSI-HIPERLAN, respectively) and has led to IEEE 802.16 (wireless-MAN) and ETSI-BRAN HIPERACCESS systems. The second is satellite technology, which is being developed by a number of vendors to provide high-capacity BWA systems with global coverage. Whereas the IEEE 802.16 and ETSI-BRAN HIPERACCESS standards have been completed and BWA products based on these standards are likely to hit the market very shortly, many of the satellite-based BWA systems are still under development and are expected to become operational in the year 2004 and beyond. This chapter describes various BWA systems and the underlying technologies.

8.1 NEED FOR BROADBAND ACCESS AND AVAILABLE TECHNOLOGY OPTIONS

As discussed in the preceding chapters, the wireless local loop (WLL) or fixed wireless access (FWA) represents a wireless-based solution for providing access to classical, narrowband, circuit-switched voice, and voice-band data services to subscribers in different market environments. However, demand for broadband communications services is rapidly increasing as private residences, small office/ home office (SOHO), small-to-medium enterprises (SMEs), and large enterprises in

Introduction to WLLs. By Raj Pandya
ISBN 0-471-45132-0

the city core add Internet access and multimedia applications to their routine communications portfolio. The increasing demand for bandwidth in the access network is being driven by a surge in demand for such services as

- Internet and Intranet access
- LAN bridging and remote LAN access
- Video-telephony and video conferencing
- Real-time video and audio (depending on economic and technical viability)
- Computer gaming
- Voice-over IP (VoIP)

According to industry projections, nearly seven out of ten people and businesses in the United States will be using the Internet by the year 2003, and broadband connections are forecasted to be the desired access method for more than 50% of these users. A significant market is therefore emerging for broadband access for delivery of these high-bit-rate services. The primary technologies that are currently deployed for providing broadband access include

- Optical fiber systems
- Digital subscriber line (DSL) over copper pairs
- Cable modems
- Broadband wireless access (BWA)

The number of new broadband access lines installed and estimated worldwide using these technologies is illustrated in Figure 8.1, where the numbers in parentheses indicate compound annual growth rates for each technology. The technical characteristics and relative economics of these broadband access methods tend to dictate their suitability for specific market environments. The market segment represented by the city core users is characterized by large groups of users concentrated in a number of closely located high-rise buildings. The high revenue potential and high demand for broadband services in the city core has supported the economics of fiber-based solutions which can be augmented by various symmetric DSL (digital subscriber line) technologies.

Immediately outside the city core with somewhat lower subscriber densities typified by small-to-medium enterprises (SMEs) and multiple dwelling units (MDUs), the technologies of choice include broadband wireless access (BWA) and symmetric DSL. Furthermore, it is estimated that over 50 million employees in the United States alone are now involved in telecommuting (remote working), indicating a growing demand for broadband connectivity needs outside the city core to support a small office/home office (SOHO) environment.

The demand for high-speed Internet access to support faster World Wide Web access and computer-based learning is also becoming very pronounced in the suburbs with their predominance of single-family homes. These are currently being

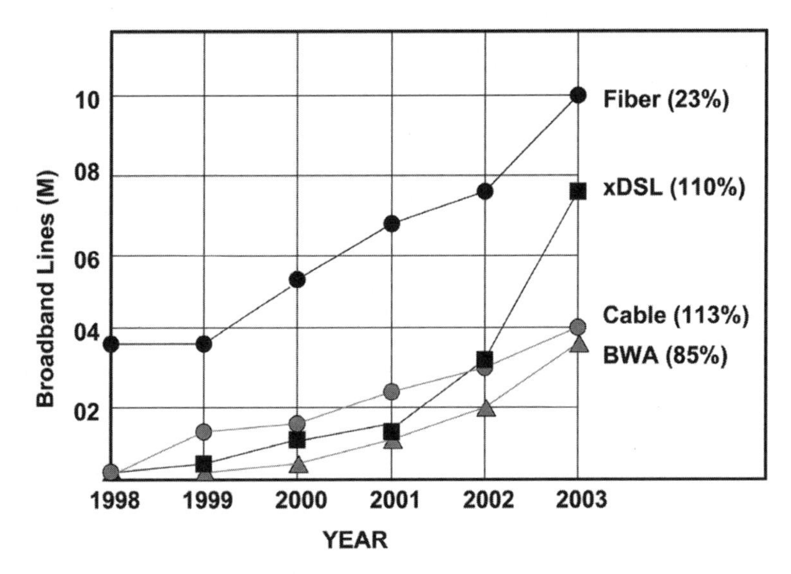

Figure 8.1. Growth in worldwide broadband access using various technologies.

met primarily by ADSL (asymmetric digital subscriber line), VDSL (very high-speed digital subscriber line), and cable modem access technologies. Recently, provision of fiber in the home is becoming more economically viable—especially fiber installation in the new suburban single-family home developments.

The BWA systems that have been deployed over the last number of years include LMDS (local multipoint distribution system), LMCS (local multipoint communication system), and MMDS (multichannel multipoint distribution system). These systems are essentially microwave point-to-multipoint distribution systems operating in a number of frequency bands ranging from 2.5 to 40 GHz and have their origin in the digital TV broadcasting and distribution systems.

The standards for systems based on wireless local area networks (WLANs) represented by 802.16 (IEEE) and HIPERACCESS (ETSI) have recently been completed. Suitable products for providing broadband wireless access based on these standards should be available shortly for delivering broadband services in some of the above market environments.

Various satellite-based broadband access systems are also an option where they can be proved in economically. Many of these systems are currently under development with initial commercial deployment expected in the 2004 time frame.

8.2 FREQUENCY BANDS AND DATA RATES FOR BWA SYSTEMS

Wireless local loop (WLL) systems, where high data rates are not a primary requirement, tend to use frequency bands below 4 GHz. However, broadband wireless access (BWA) systems may be required to support data rates from 2 Mb/s

(E1 rate) for Internet access to 155 Mb/s (STM rate) for LAN-to-LAN interconnection using point-to-point (PTP) microwave link. In order to provide higher data rates, BWA systems require allocation of larger blocks of frequencies (a few hundred MHz to more than 1 GHz) within a given frequency band. Such large blocks of frequency allocations are only available in frequency bands above 11 GHz. BWA systems operating in these high-frequency bands almost always require line-of-sight (LOS) operation, and frequencies above 20 GHz are susceptible to fading caused by rain and other types of precipitation. Table 8.1 summarizes typical frequency bands and data rates for various BWA technologies.

The management and allocation of frequency spectrum for specific applications is the responsibility of the appropriate national regulatory authority. The frequency bands indicated in Table 8.1 should be considered as representative, and the above technologies may utilize any other suitable frequency bands made available by the regulators. For example, LMDS operators in the United States were able to obtain licenses in the 24-, 28-, and 39-GHz bands and operator-specific LMDS systems at these frequency bands were provided by the manufacturers. Similarly, satellite-based systems are also available in Ku band.

At the international level, studies are underway in ITU-R for identification and sharing of frequency spectrum for fixed wireless access systems including broadband wireless access. ITU-R Recommendation F.1401 (frequency bands for fixed wireless access systems and the identification methodology) indicates that BWA applications may extend up to 70 GHz and even higher, depending on the future availability of suitable technology.

TABLE 8.1. Frequencies and Data Rates for Existing and Emerging BWA Systems

BWA System	Typical Frequency Band(s)	Nominal Data Rates (max)[a]
LMDS (USA)	28- and 31-GHz bands (FCC)	34 Mb/s
LMDS (Europe)	10, 26, and 28 GHz (CEPT)	8 Mb/s
MMDS	2.5–2.7 GHz	2 Mb/s
MVDS	40.5–42.5 GHz	6 Mb/s
IEEE 802.16	10.0–66 GHz	18.2 Mb/s
HIPERACCESS	Frequencies > 11 GHz including 26-, 28-, 32-, and 43.5-GHz bands	25 Mb/s
HIPERMAN	Frequencies < 11 GHz including 3.4 to 3.6, 3.6 to 4.2-GHz bands	25 Mb/s
Satellite systems	Ka band: 10–20 GHz (down-link) 12–30 GHz (uplink)	64 Mb/s

[a]BWA applications like video distribution and Internet access use asymmetrical channels where down-link data rates are much higher than up-link data rates.

8.3 DEPLOYMENT SCENARIO FOR BWA SYSTEMS AND TOPOLOGY OPTIONS

A typical deployment scenario for a BWA system is illustrated in Figure 8.2. The scenario consists of a single base station that serves various types of subscribers ranging from individual household to multiple dwelling units and small-to-medium enterprises. The scenario also includes an optional repeater that may be deployed to serve a cluster of customers in a localized area and/or extend the range of the base station serving area. A repeater may also be used to provide an alternate LOS path to a subscriber where the direct path has an unavoidable obstruction.

In practice, the BWA system may consist of multiple cells, each served by its base station. All the base stations in the BWA system need to be connected to the core network (PSTN, PSPDN, Internet). This can be accomplished by connecting individual base stations directly into the core network. Alternatively, the base stations may be interconnected to a *controlling* base station or base station controller (BSC) using PTP microwave links and a single PP connection from the controlling base station to the core network using appropriate interface protocols.

The deployment scenario illustrated in Figure 8.2 utilizes a centralized or star topology where the base station serves and has direct control over each major subscriber location. With the emergence of wireless LAN-based technologies for BWA, interest is growing in the deployment of more distributed or mesh topologies for BWA systems. An example scenario where mesh topography is utilized is shown in Figure 8.3.

Some of the advantages of using mesh topologies include:

- Concatenation of links can lead to increased system range.
- Greater availability of alternate paths increases the number of subscribers that can be provided with LOS paths.

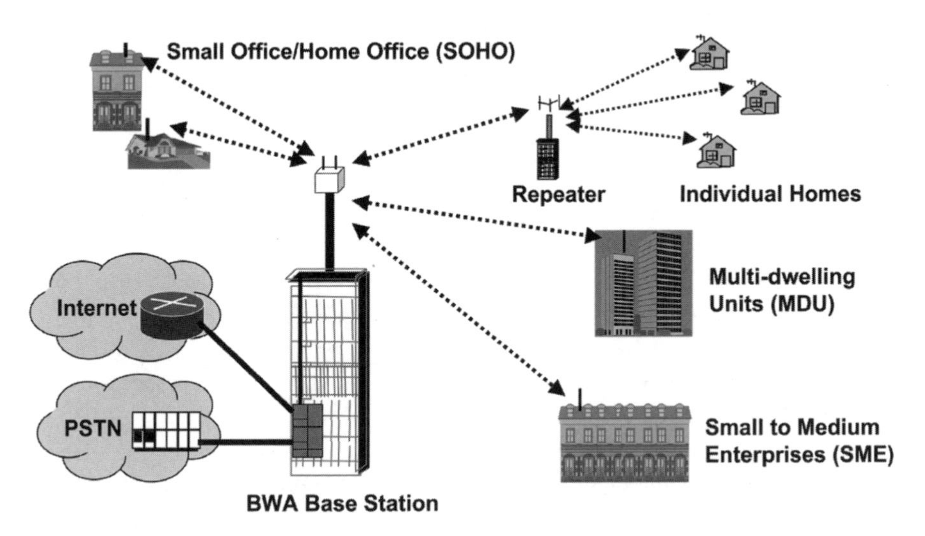

Figure 8.2. Deployment scenario for BWA system—star topology.

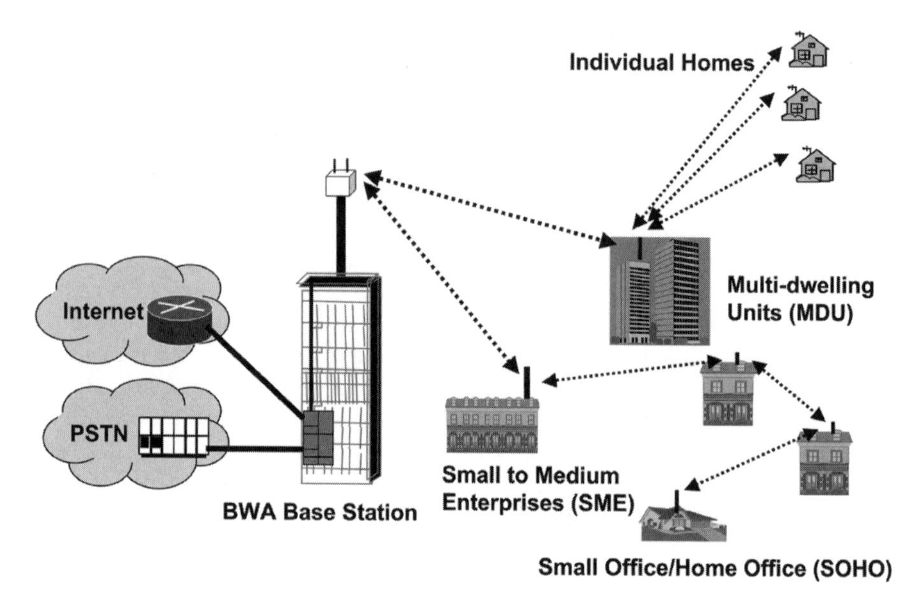

Figure 8.3. Deployment scenario for BWA system—mesh topology.

- Possibility of using smaller cells and greater frequency reuse can lead to higher system capacity.
- Availability of alternate routing paths can provide higher reliability.

The use of mesh topologies, however, poses the following technical and economic challenges that need to be addressed and factored into the choice of a deployment plan:

- Additional complexity and cost of the subscriber units which need to provide additional functionality (transmitters, receivers, antennas) for serving other adjacent subscriber units.
- Need for mounting subscriber antennas on rooftops (in order to provide visibility to other subscribers) instead of on the side of the building (to provide visibility to the base station only).
- Need for stearable subscriber antennas in order to accommodate changes in mesh topology as new subscribers are brought on board.
- Mesh networks require more complex design algorithms in order to ensure optimum frequency assignments, equitable load sharing, and proper location and beam direction of subscriber antennas.
- A distant subscriber may be subject to cumulative processing delays (decoding/recoding of signals) at the intermediate relaying subscriber units.
- Possible need for future relocation of base stations in order to accommodate changes in mesh topologies.

8.4 RADIO INTERFACES FOR BWA SYSTEMS

8.4.1 Approaches for Radio Interface Design for BWA Systems

As discussed in the preceding chapters, many of the fixed wireless access (FWA) or WLL systems operating in the lower-frequency bands are based on existing cellular mobile or cordless telecommunication radio interface standards. However, similar to conventional microwave and millimeter wave systems, most of the commercial BWA systems currently in the market deploy proprietary radio technologies.

Motivated by the technical, economic, and operational benefits of standardized radio interfaces existing in the cellular mobile, cordless telecommunication, and WLAN systems, efforts are currently underway to standardize radio interfaces for broadband wireless access. Besides the emerging 802.16 and 802.16a standards from IEEE and the HIPERACCESS and HIPERMAN standards from the BRAN (broadband radio access network) group of ETSI, ITU-R JRG 8A-9B (Joint Rapporteur Group of ITU-R Working Parties 8A and 9B) is also conducting standardization studies on various radio-related aspects of BWA.

Essentially, the radio technology options for providing wireless access for broadband services fall into the following three categories:

- In the first approach, the radio channels are allocated permanently to each subscriber leading to PTP connection between the base station and an individual subscriber. Since the base station serves a number of subscribers through separate PTP radio links, the approach is referred to as multipoint-to-multipoint (MP–MP) approach. This approach represents the direct replacement of wireline links (e.g., 2.0-Mb/s fiber or DSL links) to provide broadband services by wireless connections.
- In the second approach, a radio channel is allocated to a subscriber when a call request (from or to the subscriber) is received. The channel is returned to the pool after the call terminates. This leads to a point-to-multipoint (PTM) system, which provides better spectrum usage and lower system cost. This approach, however, is better suited for providing narrowband PSTN services to a large subscriber base (e.g., cellular mobile and WLL systems).
- In the third approach, a high bandwidth radio channel is available to all the subscribers (served by a base station) at all times. An appropriate amount of bandwidth is made available to individual subscribers when traffic, in the form of ATM cells or IP packets, needs to be transmitted (from or to the subscriber). This approach provides the most efficient spectrum usage compared to the previous two methods, provided that the overhead bits are kept within reasonable limits. The approach is best suited for BWA systems based on wireless LAN concepts.

8.4.2 Multiple Point-to-Point (MPP) Approach

This approach aligns with the classical PSTN approach of providing protocol and application independent transmission pipes to the subscribers. It was adopted by early BWA systems that essentially tried to replace wireline transport (like fiber, xDSL, cable) by wireless connections to provide broadband services. The classical approach of multipoint-to-multipoint (MP–MP) connections as one-to-one replacement of wireline connections is illustrated in Figure 8.4. Channel bit rates of multiples of 64 kb/s (n × 64), 1.55 Mb/s (T1 rate), or 2.08 Mb/s (E1 rate) are available to the users, and the channels can be assigned on a demand basis. Some systems can also provide higher bit rates up to 4 × E1 per customer. Although either FDMA/FDD or TDMA/FDD methods can be used, most early systems were designed using FDMA/FDD methods.

The point-to-point nature of channel usage and assignment in these systems provides the flexibility to alter the modulation scheme used on the channel (on a call-by-call basis), depending on the prevailing propagation conditions at the user location. High-level modulation schemes like 64-QAM can be deployed to achieve very high data rates on the channel. However, use of different modulation schemes on an individual channel on a call-by-call basis introduces different interference sensitivity requirements at the receiver, thereby complicating the radio network planning.

The usage of radio spectrum in the MP–MP approach is not very efficient insofar as it does not provide any multiplexing or concentration in the radio network. Concentration for voice services can be provided by use of V5.2 interface between the base station and the PSTN. However, carriage of asymmetric and bursty data traffic generated by Internet services or interactive data applications cannot be handled very efficiently by the MP–MP approach. The MP–MP approach is therefore best suited for carriage of symmetric, constant bit rate traffic.

8.4.3 ATM and IP Approach

In the last number of years the focus of BWA has shifted toward offering a wide portfolio of services that include voice, high-speed data, Internet access, interactive

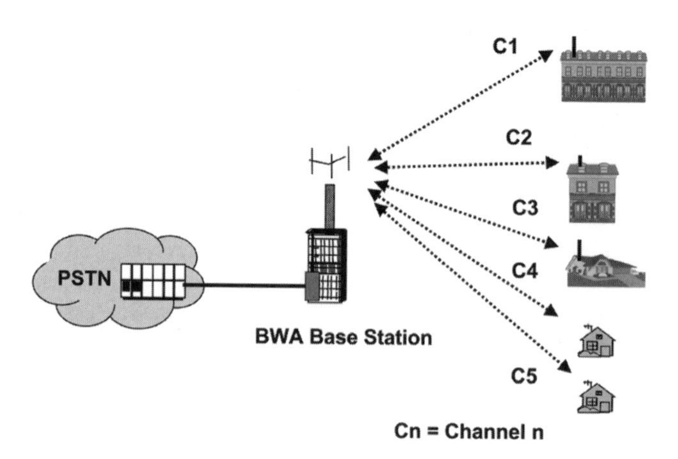

Figure 8.4. Multipoint-to-multipoint (MP–MP) connections for BWA.

services, and multimedia services while ensuring the most efficient usage of radio spectrum. This has led to radio interface designs that utilize ATM or IP concepts for transfer of user information across the radio interface. In this type of radio interface, a pool of bandwidth within a radio coverage area is created, which is then shared between different subscribers using statistical multiplexing. The radio channel capacity is assigned only when user traffic in the form of ATM cells or IP packets are to be transported across the channel. The bandwidth and service quality requirements for each traffic stream are obtained from the cell/packet header. The statistical multiplexing of different traffic types on the radio channel is illustrated in Figure 8.5.

In principle, all of the bandwidth may be used by one user, or shared according to a suitable multiple access technique among all of the users demanding bandwidth. In practice, depending on the application, the user may demand different amounts of bandwidth in the up-link and down-link directions. The different traffic types (1, 2, 3, 4) shown in Figure 8.5 may result from the following applications:

1. A voice call using a 64-kb/s channel in both down-link and up-link channels
2. Down-link bursts representing a single 2-Mb/s MPEG encoded video stream being downloaded from the network to the user terminal
3. Irregular down-link data bursts representing a World Wide Web page being downloaded to a PC running a web browser application, and the up-link mouse pointer data or TCP/IP acknowledgments
4. Long up-link data bursts representing segments from a large data file segmented to accommodate the short segments of isochronous traffic type 1

Fundamental to the operation of the scheme is the fact that an application which only has requirements for low average bit rate uses the entire available bandwidth during the short interval while it has access to the radio channel. The higher the maximum available channel bit rate, the more efficient the statistical multiplexing becomes.

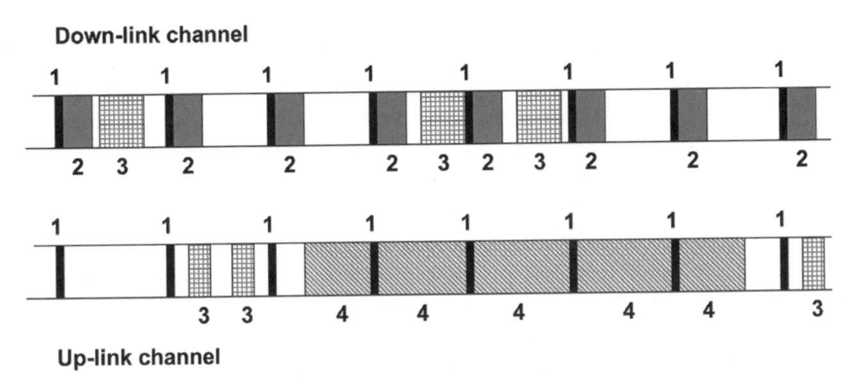

Figure 8.5. Statistical multiplexing of different traffic types on a TDMA/FDD radio channel.

A factor that needs to be considered in this approach is the different priorities that different data types will impose. For example, in order to maintain the quality of service for speech and video information transfer, it is essential that such traffic is not subject to variable delay. This may require that such traffic receive highest priority for transmission over the radio channel. Web data and file transfers, on the other hand, are not time critical and can be delayed and, if necessary, queued at the base station or user terminal while the radio channel is busy. The radio interface design needs to ensure that different traffic streams can be mixed efficiently and can achieve high channel utilization and provide all the different users acceptable service quality for their applications.

Though Figure 8.5 illustrates the use of a frequency division duplex (FDD) structure, a time division duplex (TDD) is equally suitable for this approach. In fact, TDD is better suited for applications requiring asymmetric channels whose relative bandwidths can be easily altered by changing the number of time slots assigned to the up-link and down-link directions.

The BWA systems like IEEE 802.16 and ETSI-BRAN HIPERACCESS, which are based on wireless LAN concepts, are standardizing radio interfaces based on this approach. Furthermore, ITU-R JRG 8A-9B is also developing Recommendations on radio local area network (RLAN) technologies that efficiently convey ATM cells or IP packets.

8.5 BWA SYSTEMS BASED ON MULTIPOINT DISTRIBUTION CONCEPTS

8.5.1 Introduction

BWA systems based on multipoint distribution concepts resulted from the convergence of broadcasting and communication technologies in the microwave frequency bands. In the United States, the 2-GHz band was initially assigned for broadcast of analog TV signals under the name of multichannel multipoint distribution systems (MMDSs). These were intended as a *wireless cable* alternative to wired cable networks for distribution of analog television signals. Analog MMDS systems are widely deployed in the United States, where the 2-GHz spectrum allocation allows for 336 channels, each having 6-MHz bandwidth. The nominal reach is in the 40-km range, depending on the antenna height.

However, the potential for utilizing an adaptation of the technology for two-way and interactive digital data services was soon recognized. The digital data services could include broadcast of multimedia entertainment (digital TV), Internet access, and on-demand video services. In 1998–1999 the FCC conducted auctions for frequencies in the 28-GHz and 40-GHz bands for this purpose, and the resulting systems are called local multipoint distribution systems (LMDSs).

Before the allocation of the above frequency bands resulting in the advent of LMDS in the United States, a 25.5 to 27.5-GHz band was licensed in Canada for a terrestrial microwave system to support pay TV, Internet access, video conferencing, and other multimedia services. The system in Canada was designated as local multipoint communication system (LMCS) and, in a sense, represents a precursor to LMDS.

Multipoint video distribution system (MVDS) and cellular access to broadband services and interactive television (CABSINET) are two European systems in this category for use in the 40-GHz band, with main focus on distribution of television and video signals. DAVIC (digital audio–visual council) has also developed physical layer specifications for carriage of video signals over radio frequencies ranging up to 10 GHz. The scope of the initial DAVIC specification is to support TV distribution and video on demand (VoD) applications. Subsequent DAVIC specifications are intended to support bi-directional digital transmissions above 10-GHz bands with capability for transport of both MPEG-2 video and ATM cells in the down-link.

8.5.2 Local Multipoint Distribution Systems (LMDSs)

LMDSs combine high-capacity radio-based communication capability with broadcasting in an interactive manner in the microwave frequency bands. The possibility of adapting the MMDS concept of full service broadcast network into an interactive communications network by adding communications channel for a return path coincided with the tremendous growth in the Internet and high-bit-rate data services. As opposed to MMDS broadcast systems, which operate in relatively lower frequency bands and have large coverage radius, the higher-frequency bands used for LMDS restrict their coverage range and are best suited for local distribution. Furthermore, because of the strict line-of-sight operation imposed by the higher frequency bands, most LMDS are deployed outside the city core for serving small-to-medium enterprises (SMEs), multiple dwelling units (MDUs), and small office/home office (SOHO) users.

In the United States, 1.3 GHz in the 28- and 31-GHz band has been assigned for LMDS by the FCC. The frequency assignments are illustrated in Figure 8.6. European countries are allocating frequencies in a number of different bands, with the primary high-capacity band currently being 40.5–42.5 GHz, which is being extended to 43.5 GHz. This high-frequency band is likely to be shared by multiple operators competing in the same geographic service area. However, since frequency band allocations is a national responsibility, licensing and deployment of LMDS in Europe is also likely to take place in many other frequency bands including the 24-, 26-, and 28-GHz CEPT frequency bands.

LMDS can be deployed in a multicell configuration in order to provide required coverage. Separate base stations with co-located transmitter/receiver are deployed at each cell site. One of the base stations serves as a coordination center for the entire franchise area and connects the LMDS cells to the public networks. The base stations can be interconnected using either fiber links or short hop point-to-point microwave radio links. Figure 8.2 may be considered as a typical architecture for a single cell in an LMDS implementation.

As mentioned earlier, operation in the typical LMDS frequency bands restricts the operation to line of sigh links and is also subject to severe fading or attenuation

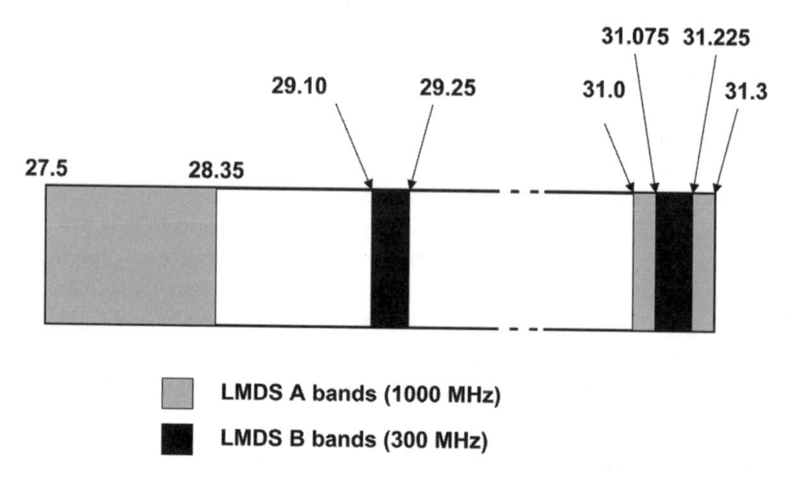

Figure 8.6. LMDS frequencies allocated by the FCC in the United States.

due to precipitation. Vegetation and foliage in the vicinity of the subscriber location can be another contributing factor to severe signal attenuation requiring close attention to placement of the antenna at the subscriber site.

These effects limit the reliable range of LMDS to 3–5 km, depending on the climate zone and operating frequency. Table 8.2 illustrates the sensitivity to frequency and climate zones on the coverage radius of a BWA system based on multipoint distribution technology. ITU-R has developed tables and maps for different world climate zones ranging from climate zone A (very low precipitation profile) to climate zone Q (very high precipitation profile). Climate zones E and K represent intermediate precipitation profiles where precipitation levels in zone K are approximately twice that in zone E.

Because of the line-of-sight (LOS) restrictions, it is difficult to ensure adequate coverage to all the potential subscribers to be served by a cell site. About 95%

TABLE 8.2. Coverage Sensitivity to Operating Frequency and Climate Zone

Frequency (GHz)	ITU-R Climate Zone	Coverage (km)
26	E	5.5
26	K	3.5
28	E	5.1
28	K	3.2
32	E	4.1
32	K	2.6
43.5	E	3.3
43.5	K	2.2

coverage is expected for providing broadband services in a public offering. In city core the LOS coverage may be as low as 50%, and adequate measures are required to overcome the coverage problem. Besides locating the cell site antennas at a suitable height, repeating stations or reflectors may be deployed (at extra cost) to reach subscribers that are in a shadow. Furthermore, providing overlapping cells in a multicell implementation can allow a subscriber to be served by a base station in a neighboring cell which may be able to provide a LOS path. In the worst case, a number of suitably located additional base stations may be the only solutions.

Except for some general guidelines in terms of service offerings and frequency bands, there are no regionally or internationally agreed standards for LMDS. Different manufacturers have developed their own proprietary systems. Systems using both FDMA and TDMA access methods are available. The FDMA-based systems are preferred where subscribers want permanent links so that a FDMA frequency channel can be assigned permanently to the subscriber. No medium access control is required in such applications. TDMA-based systems are generally used where subscribers request access to bandwidth on a demand basis. The number of TDMA time slots allocated to the subscriber at any one time can be matched to the amount of bandwidth required for the application.

Many LMDS implementations offer the flexibility of utilizing alternate modulation schemes that are preprogrammed and can be selected in order to maximize spectrum efficiency under prevailing interference conditions. The bandwidth required to provide a fixed 2.08-Mb/s link (E1 link) under different modulation schemes used in LMDS is illustrated in Table 8.3.

As illustrated in Table 8.3, the higher-level modulation schemes like 16-QAM and 64-QAM offer much greater efficiency in spectrum utilization. However, deployment of these higher-level modulation methods require higher signal-to-noise ratio (SNR) at the receiver, thereby adversely affecting the system range. Furthermore, these schemes are also more susceptible to interference, resulting in the potential need for higher-frequency reuse ratios where cells are closely clustered. Overall user bandwidth may therefore be reduced. The higher modulation schemes

TABLE 8.3. Bandwidth Required for Different LMDS Modulation Scheme

Modulation Scheme	Required Bandwidth (MHz)
Binary phase shift keying (BPSK)	2.8
Differential quadrature PSK (DQPSK)	1.4
Quadrature PSK (QPSK)	1.4
8-state PSK (8-PSK)	0.8
4-level quadrature amplitude modulation (4-QAM)	1.4
16-level QAM (16-QAM)	0.6
64-level QAM (64-QAM)	0.4

are generally selected where the distance from the base station to the subscriber units is lower where adequate SNR is expected to be available.

In terms of performance and capacity, availability objectives in the range of 99.99% to 99.999% are generally quoted by LMDS service providers. These objectives may be further improved with the expected progress in radio technology and the emerging need to provide performance comparable to fiber-based broadband access. Consider a typical LMDS implementation with 2 × 100 MHz of duplex frequency allocation and which uses a 6-sector antenna with a frequency reuse plan of 2. The range of capacity available (in terms of number of 2.08-Mb/s E1 links) in a cell for different modulation methods is illustrated in Table 8.4.

Similar to WLL systems, LMDS radio equipment consists of the transmitter/receiver at the base station and remote subscriber units which contain an outdoor unit (ODU) and an indoor unit (IDU). The primary difference between the LMDS and WLL radio equipment is (a) the higher frequencies deployed for LMDS which affect the radio design and (b) the increased functionality of the IDU at the remote subscriber station which needs to support a wider range of devices which may include a LAN, a PBX, a video terminal, and various n × 64 and T1/E1 links. The antenna at the base station typically uses a four- or six-sector arrangement with a frequency reuse factor of two.

Besides the need for meeting the functionality, performance, regulatory, and environmental requirements, the general criteria for the design and deployment of the LMDS radio equipment include

- Cost effectiveness
- Freedom from maintenance
- Capability for quick and easy installation
- Capability for easy upgrades
- Scalability
- Flexibility

TABLE 8.4. LMDS Capacity (E1 Links/Cell) as a Function of Modulation Method

Modulation Method	Required BW (MHz/E1 Link)	Capacity (E1 Links/Cell)
BPSK	2.8	107
DQPSK	1.4	214
QPSK	1.4	214
8-PSK	0.8	375
4-QAM	1.4	214
16-QAM	0.6	500
64-QAM	0.4	750

The radio equipment (ODU) at the remote subscriber unit consists of transmitter and receiver circuits (amplifiers, band-pass filters, automatic gain control, etc.), a local oscillator and mixers, a diplexer connected to the antenna, and a cable interface for connection to the IDU. For small business and residential markets, the cost of the subscriber units is a primary factor and simple radio designs with *minimal radio intelligence* are required. Advent of software controlled radio technology is likely to play an important role in cost reductions and greater penetrations of LMDS in these markets.

8.5.3 Multipoint Video Distribution System (MVDS)

MVDS is a system similar to LMDS and is a microwave-based wireless distribution system. However, unlike LMDS, which has emerged as a system supporting a broad range of broadband services, the primary focus of MVDS has remained on video distribution. Furthermore, MVDS has a better defined standard than LMDS and is based on the digital video broadcasting (DVB) standard developed by ETSI, and it utilizes the 40-GHz band assigned by the CEPT (Conference of European Post and Telecommunications) in Europe. Some of the key design characteristics of MVDS are summarized below:

- The frequency band 40.5–42.5 GHz has been harmonized within the CEPT for DVB, which is the basis for MVDS.
- Out of the available 2-GHz spectrum, 100 MHz is used for the up-link or return path to allow the user to interact with the video service provider (to select the nature and bit rate of the video download).
- The system is suitable for use in various transmission bands other than a 40-GHz band.
- It is compatible with MPEG-2 (Moving Pictures Expert Group 2)-coded TV services and provides transmission capability synchronous with packet multiplexing.
- The multiplexing flexibility allows the use of transmission capacity for a variety of TV service configurations including sound and data services.
- Since MVDS services using high-frequency microwave transmissions are more susceptible to power limitations (rather than noise and interference), system robustness rather than spectral efficiency is the primary design criterion.
- To achieve very high power efficiency while maintaining good spectral efficiency, the system deploys QPSK modulation and concatenation of convolution and R–S (Reed–Solomon) channel coding.
- The system is designed to support a single carrier per MVDS transmitter using time division multiplex (TDM) as well as for multicarrier frequency division multiplex (FDM).

A functional representation of an MVDS down-link channel is shown in Figure 8.7. The video, audio, and data components are source-coded and

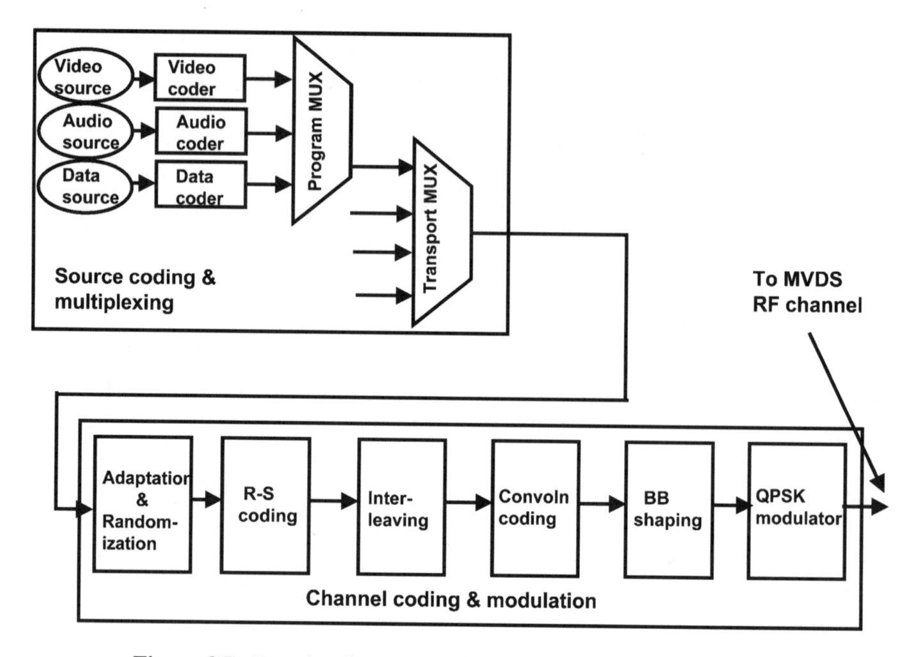

Figure 8.7. Functional representation of MVDS down-link channel.

multiplexed and constitute the information stream for an individual program. Multiple program streams are then combined at the MPEG-2 transport multiplexer and submitted to the MVDS channel adapter for channel coding and QPSK modulation.

The adaptation and randomization function ensures full conformance with the MPEG-2 packet formats and ITU-R radio regulations for adequate binary transitions, respectively. The base-band (BB) shaping function consists of a square-root raised cosine filter.

8.5.4 Cellular Access to Broadband Services and Interactive Television (CABSINET)

CABSINET is a European TV distribution system that utilizes a two-layer architecture to provide last-mile TV distribution service in both urban and rural market environments. The first layer is similar to the MVDS and utilizes the 40-GHz band with LOS paths providing a coverage radius of 1–3 km. The second layer uses a local repeater which down-converts the 40-GHz signal to a lower frequency in the 5.7-GHz band for distribution in a limited geographic area with a range of 50–500 m. The use of the lower frequency permits the signal to penetrate through building walls and makes the system suitable for delivery of TV signals inside homes with simple set top receivers. The two-layer architecture for CABSINET is illustrated in Figure 8.8.

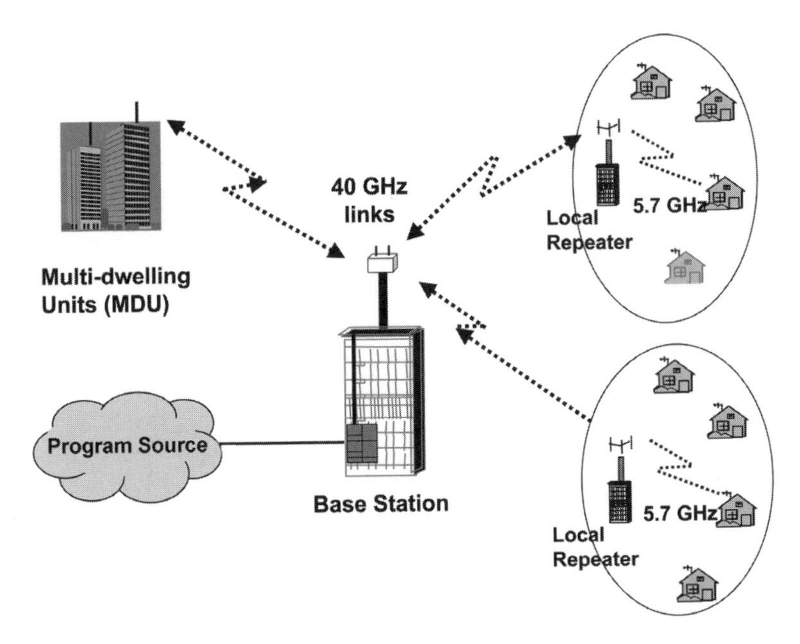

Figure 8.8. Two-layer architecture for CABSINET system.

The transmissions in the macro cells utilize QPSK modulation and can encode four 6-Mb/s programs within each 39-MHz FDMA/TDMA channel. The base stations use 90-degree sectored antennas. The micro-cell transmissions operate with a bandwidth of 150 MHz with coded orthogonal frequency division multiplex (OFDM) modulation to encode one 8-Mb/s program into an 8-MHz TDMA or DS-CDMA channel.

As indicated in Figure 8.8, there are two access options for the users. They may access the service either directly at the macro-cell level over the 40-GHz link or through the micro cell using the 5.7-GHz links. The radio links are designed for asymmetric operation with the flexibility to alter the up-link bandwidth based on users' requirements for video and data applications. The micro cells can be configured as a wireless LAN in a mesh configuration or in a star configuration, depending on the subscriber distribution within the coverage of the micro cell.

8.5.5 Examples of Commercially Available BWA Systems Based on Multipoint Distribution

There are a number of commercial MMDS and LMDS systems that have been implemented throughout the world. All of these systems are based on multipoint distribution concepts using proprietary technologies developed by a number of manufacturers and operate in a range of MMDS/LMDS frequency bands. Key features of some selected BWA systems are summarized in Table 8.5.

TABLE 8.5. Key Features of Some Selected Commercial BWA Systems

System Parameter	AIReach^{TM} (Hughes Network Systems)	7390 LMDS (Alcatel)	Beewip^{TM} (Ericsson)	OnDemand^{TM} (Lucent Technologies)
Frequency bands (GHz)	24–48; various bands	28–43; various bands	3.4–3.6	24–38; various bands
Multiple access	TDMA	TDMA	FH-CDMA	FDMA
Duplexing	FDD	FDD	FDD	TDD/FDD
Channel BW (MHz)	12.5, 14.0	14, 28 (DL) 3.5, 7.0 (UL)	3.0	25
Gross bit rate (Mb/s)	15–45	20, 40 (DL) 5, 10 (UL)	1, 2, 3	41
Modulation	QPSK, 16-, 64-QAM (adaptive)	QPSK (DL) DQPSK(UL)	GFSK (1, 2, 3 bits)	4-, 16-, 64-QAM (adaptive)
Nominal range (km)	0.94–7.34	5.0	2.5–10.0	4.0
Services	nxT1/E1, 10/100 base T, ATM, IP	nx T1/E1, 10 base T, ATM, IP	10/100 base T, IP, ATM	Voice, data, IP, LAN interconnection, ATM

Besides the systems in Table 8.5, there are many other commercial LMDS-based BWA systems that have been in the market over the last number of years and have been implemented in various countries. Examples include systems like Reunion™ (Nortel Networks), WALKair™ (Siemens), SpectraPoint™ (Motorola), WATMnet™ (NEC), and WAVEACCESS™ (Lucent Technologies). Many of these systems deploy multiple modulation schemes in an adaptive manner to provide best trade-off between signal quality and system range. They also provide capabilities for operation in a number of frequency bands in order to accommodate operators who have licenses in different bands. The majority of these systems support asymmetric operation.

8.6 BWA SYSTEMS BASED ON WIRELESS LAN CONCEPTS

8.6.1 Introduction

The popularity and the market potential for wireless local area networks (WLANs) are driven by such applications as inventory control in warehouses and stores, point-of-sale terminals, rental car check-ins, and real-time patient record updates in hospitals. A very rapidly emerging application of WLANs is computer wireless networks that provide local mobility and Internet connectivity to laptop computers fitted with PCMCIA cards in homes, offices, and public places like airports. LAN-to-LAN interconnection is another emerging application for WLAN technology.

In order that the available frequency band can be utilized efficiently and fairly by all the users and to ensure interoperability of terminals and equipment from multiple vendors, two wireless LAN standards have emerged. The North American standard for WLAN developed by the IEEE is called 802.11, and the European standard developed by ETSI is called HIPERLAN (high-performance LAN). IEEE 802.11 uses about 85 MHz in the 2.4-GHz band, which is part of the unlicensed ISM (Industrial, Scientific, and Medical) band of frequencies in the United States. HIPERLAN operates in the 5-GHz band with a bandwidth of 150 MHz.

In recent years it has been recognized that the WLAN technology can be extended to provide broadband interactive services to subscribers in the metropolitan area network (MAN) market, which requires wider coverage range and higher data rates. Besides leveraging the economies of scale from the existing WLAN market, this approach has the advantage that it is well-suited to support Internet access and interactive data services because of its inherent packet-based transport and protocol design. The standards that have been completed for this purpose include IEEE 802.16 for 10- to 66-GHz frequency bands and ETSI-BRAN HIPERACCESS for frequency bands above 11 GHz. Efforts are currently underway in the two standards development organizations to harmonize, as far as possible, the two standards and their future extensions.

IEEE 802.16a and ETSI-BRAN HIPERMAN are two additional standards for operation in the 2- to 11-GHz (licensed and unlicensed) bands that have actively been worked on in these standards development organizations and are expected to be completed shortly. The need for these standards in lower frequency bands is motivated by support of non-line-of-sight (NLOS) applications that can satisfy the needs of the residential market with reduced cost and equipment complexity.

Both 802.16 and HIPERACCESS standards are being specified to support the following services, though a specific implementation may only support a subset of these:

- Legacy time division multiplex (TDM) voice and data
- IP connectivity and Internet access
- Packetized voice-over IP (VoIP)
- Video telephony and conferencing
- Video on demand (VoD)
- Computer gaming
- Remote LAN access

Topology for 802.16 or HIPERACCESS wireless metropolitan area network (WMAN) is illustrated in Figure 8.9. These WMANs differ significantly from 802.11 and HIPERLAN wireless local area networks (WLANs) in terms of topology and the functionality of their components. WMANs are characterized by the following:

- Base stations (BSs) are connected to public networks.
- A subscriber station (SS) served by a BS typically serves a building (business or residence).

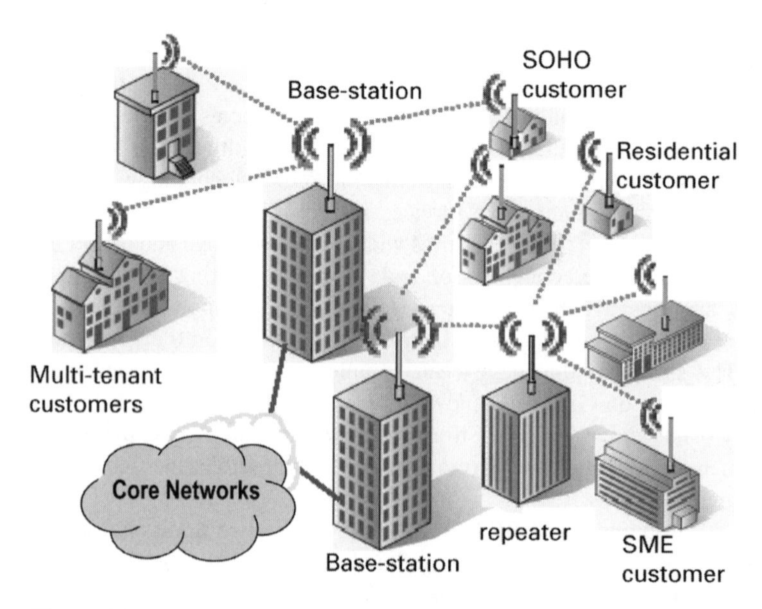

Figure 8.9. Topology for a wireless metropolitan area network (WMAN).

- An SS provides last/first-mile access to public networks.
- Multiple services, with different QoS priority, are supported.
- Both contention-free and contention-based access is provided.

A WMAN supports many more users, with much higher data rates, at much longer distances than a WLAN.

8.6.2 IEEE 802.16 Broadband Wireless Access Standard

8.6.2.1 Scope of IEEE 802.16 Standard The IEEE 802.16 standard primarily consists of the specification for the radio interface, which provides the underpinnings for the design and deployment of metropolitan area networks to provide broadband wireless access in the 10- to 66-GHz licensed frequency bands. The 10- to 66-GHz radio interface is designated as *WirelessMAN-SC*, where SC indicates single carrier operation.

The radio interface specified by the IEEE 802.16 standard includes the medium access control (MAC) layer and physical (PHY) layer of fixed point-to-multipoint (PTM) broadband wireless access systems supporting multiple services and traffic types. The MAC layer is capable of supporting multiple PHY layer specifications optimized for the frequency bands of the application. The standard includes a particular PHY layer specification broadly applicable to systems operating between 10 and 66 GHz. The scope of the radio interface specification is illustrated in Figure 8.10.

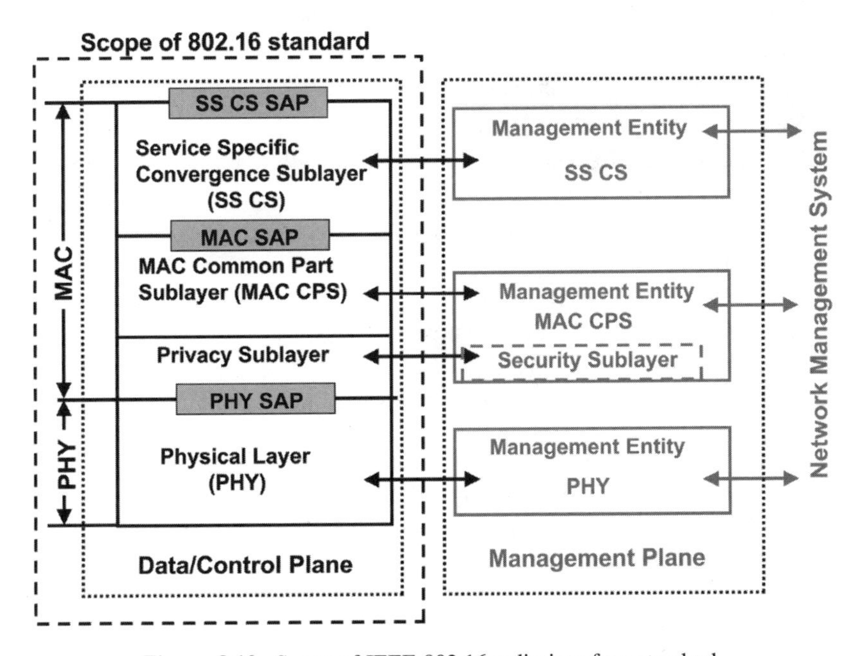

Figure 8.10. Scope of IEEE 802.16 radio interface standard.

Though the underlying intent of the standard is to promote interoperability between multi-vendor equipment and lower operator costs, the standard also provides built in features and tools for vendor differentiation. Examples include the adaptive burst profile feature for optimizing the efficiency of transport at the PHY layer and provision of a set of tools for the base station to implement efficient scheduling.

8.6.2.2 IEEE 802.16 PHY Layer Features

The 802.16 PHY layer specification is designed for the 10- to 66-GHz band, which assumes line-of-sight propagation and minimal multipath fading over the transmission path. For this reason a single carrier (SC) modulation was selected where on the down-link the base station transmits a time division multiplex (TDM) signal, with individual subscriber stations assigned time slots in a serial manner. Subscriber stations utilize the TDMA method to access the up-link channel. A burst design that can operate with either FDD or TDD is implemented. The burst profiles (on TDD or FDD) are adaptive in that different modulation and channel coding methods can be assigned dynamically on a burst-by-burst basis, depending on the channel conditions, to the individual subscriber stations. Key features of the 802.16 PHY layer include the following:

- Supports wide channels (20, 25, or 28 MHz).
- Provides high capacity (high bit rate) on both down-link and up-link.
- Uses TDM for down-link and TDMA for up-link (SS assigned variable length time slots on up-link).
- Utilizes adaptive burst profiles on down-link and up-link.
- Supports FDD and TDD.
- Dynamic asymmetry is provided under TDD operation.
- Supports half-duplex terminals.
- Modulation and FEC can be assigned dynamically on a burst-by-burst basis, depending on link conditions.
- Modulation methods include SC-QAM (gray-coded), QPSK, 16-QAM, and 64-QAM.
- FEC methods include Reed–Solomon (R–S), R–S concatenated with BCC codes, and optional TPCs (turbo product codes).
- Uses a frame length of 1 ms with allocation based on physical slots (PS) where PS = 4 modulation symbols = 1, 2, or 3 bytes, depending on modulation scheme.

The PHY layer design provides the flexibility to the equipment manufacturers for choosing the channel size based on spectrum requirements. Different modulation schemes then result in different bit rates. Table 8.6 illustrates the relationship between channel size, modulation methods, and bit rates.

8.6.2.3 IEEE 802.16 MAC Layer Features

The 802.16 MAC protocol is designed for point-to-multipoint (PTM) broadband wireless applications and

TABLE 8.6. Example of Relationship Between Channel Size, Modulation, and Bit Rates

Channel (MHz)	Symbols (MS/s)	Mb/s (QPSK)	Mb/s (16-QAM)	Mb/s (64-QAM)
20	16	32	64	96
25	20	40	80	120
28	22.4	44.8	89.6	134.4

addresses the need for very high bit rates in both the down-link and up-link directions. It provides for access and bandwidth allocation algorithms that can support a large number of terminals per channel including terminals shared by multiple end users. The 802.16 MAC layer protocol is designed to meet the following requirements:

- Achieve maximum spectrum efficiency in a wireless metropolitan area network environment in the 10- to 66-GHz frequency bands.
- Support PTM access.
- Support continuous and bursty traffic in contention-oriented mode.
- Achieve a balance between stability of contention-less operation and efficiency of contention-based operation.
- Support very high bit rates in both down-link and up-link directions.
- Provide flexible QoS offerings (CBR, rtVBR, nrt-VBR, BE with granularity within classes).
- Interface efficiently into Ethernet, IPv4, IPv6, and ATM networks.
- Support terminal sharing by multiple end users.
- Support large subscriber base per base station.
- Provide for secure access and privacy on *on-the-air* transmissions.
- Support PHY layer alternatives including single carrier, FDD/TDD, adaptive modulation, and FEC.

Some other characteristics of the IEEE 802.16 MAC layer specifications are as follows:

MAC addressing is based on

- 48-bit IEEE MAC address for the subscriber stations,
- 48-bit ID for the base stations (not an IEEE MAC address),
- 24-bit operator indicator, and
- 16-bit connection ID (CID).

MAC PDU transmission takes place in bursts at PHY layer, where:

- A PHY layer burst may contain multiple concatenated MAC PDUs.
- A PHY layer burst may contain multiple FEC blocks.
- MAC PDUs may span FEC block boundaries.

Down-link transmissions consist of two types of bursts: TDM and TDMA. In addition:

- All bursts are identified by a DIUC (down-link interval usage code).
- TDMA bursts have resync preamble for more flexible scheduling.
- Each terminal listens to all bursts at its operational DIUC (or a DIUC with a stronger signal) except when the terminal is in a transmit mode.
- Each burst may contain data for several terminals.
- A subscriber station needs to recognize the PDUs with known CIDs.
- A DL-MAP defines usage of down-link and contains carrier-specific data.

Up-link transmissions are invited transmissions, where:

- Transmissions in contention slots are used for bandwidth requests and the contentions are resolved using *truncated exponential back-off (TEB)*.
- Transmissions in initial ranging slots contain *ranging requests (RAN-REQ)* with TEB contention resolution.
- Up-link bursts are defined by up-link interval usage codes (UIUCs).
- A UL-MAP message allocates up-link transmissions.
- All up-link transmissions have a synchronization preamble.
- All data from a single subscriber station is concatenated into a single PHY burst.

Up-link service classes include the following service flow (SF) characteristics:

- Unsolicited grant services (UGSs) for constant bit rate (CBR) service flows such as T1/E1.
- Real-time polling services (rtPSs) for real-time variable bit rate (rtVBR) service flows such as MPEG-2 video.
- Non-real-time polling services (nrtPSs) for non-real-time service flows with better than best effort (BE) services like bandwidth-intensive file transfers.
- Best effort (BE) for best effort (background) traffic.

Request/grant scheme for up-link transmissions has the following properties:

- It is self-correcting with no acknowledgment and all errors handled through periodic aggregate requests.

- Bandwidth requests are always per connection.
- Grants are either per connection (GPC) or per subscriber station (GPSS).
- GPC is more suitable for a SS with many end users and requires more intelligent SS.
- GPSS is suitable for simpler SS with few users.
- Grants (specified as duration) are carried in the UL-MAP messages.
- Subscriber station needs to convert the allocated time to data bits using UIUC information.

8.6.2.4 IEEE 802.16 Privacy Sublayer (Authentication and Encryption) The 802.16 specification provides for privacy protection by encryption of over-the-air transmissions, as well as for authentication and authorization of subscriber stations to prevent theft of service and cloning of terminals. For the purpose of subscriber station (SS) authentication, each SS contains a factory-installed, manufacturer-issued X.509 digital certificate as well as a certificate identifying the SS manufacturer. These certificates are used to establish a binding between the 48-bit MAC address of the SS and the public RSA key. These certificates are sent by the SS to the network (i.e., BS) in the authentication request and authentication information messages. The network verifies the identity of the SS by checking the X.509 and the manufacturer's certificates. If the authentication is successful, the BS sends the authentication reply message containing an authentication key (AK) encrypted by the SS's public key and is used to encrypt future transmissions.

IEEE 802.16 privacy protocol is based on suitable enhancements to the privacy key management (PKM) protocol in order to operate efficiently with the 802.16 MAC layer protocol and to accommodate stronger encryption algorithms like the advanced encryption standard. The PKM protocol uses X.509 digital certificates with RSA public key encryption for SS authentication and authorization key exchange. The data encryption standard (DES) running in the cyber block chaining (CBC) mode with 56-bit keys is deployed for traffic encryption with a CBC initializing vector based on frame numbers. Critical MAC management messages like *call setup* are also authenticated using PKM.

8.6.2.5 IEEE 802.16 Service Specific Convergence Sublayer The service–specific convergence sublayer provides mapping services to and from the 802.16 MAC connections. The two main convergence layers are the ATM convergence sublayer (ATM CS) and the packet convergence sublayer (packet CS).

ATM CS provides support for

- Virtual path (VP)-switched connections
- Virtual channel (VC)-switched connections
- End-to-end signaling of dynamically created connections
- ATM header suppression for better transport efficiency
- Full range of QoS

Packet CS similarly provides support for

- Ethernet, IPv4, and IPv6 (initially)
- Payload header suppression (generic as well as IP-specific)
- A range of QoS options
- Possibility of future support for PPP, MPLS, etc.

As mentioned earlier, the IEEE 802.16 wireless MAN standard for the 10- to 66-GHz standard will be supplemented by shortly to be completed 802.16a specification addressing the 2- to 11-GHz band. The design of the PHY layer for 802.16a is driven by non-line-of-sight (NLOS) operations targeted for a point-to-multipoint distribution of broadband services to the residential market. In order to counter potential multipath fading in the lower frequency bands, 802.16 intends to specify three different radio interfaces designated as

- Wireless MAN-SC2, which is similar to wireless MAN-SC but at 2–11 GHz
- Wireless MAN-OFDM, which uses orthogonal frequency division multiplexing with a 256-point transform and is mandatory for unlicensed bands
- Wireless MAN-OFDMA, which uses OFDM with a 2048-point transform.

8.6.3 ETSI-BRAN HIPERACCESS Broadband Wireless Access Standard

8.6.3.1 Scope of HIPERACCESS Standard HIPERACCESS is one of a series of specifications developed by the broadband radio access network (BRAN) project under ETSI. The primary aim of the BRAN project is to develop standards for the new generation of service-independent broadband radio access networks and systems using licensed and license-free spectrum. As shown in Table 8.7, BRAN embraces specifications that include: HIPERLAN/2, a high-performance radio LAN standard with centralized control; HIPERACCESS, a fixed point-to-multipoint wireless metropolitan area network; and HIPERLINK, a low-cost, high-capacity, short-range point-to-point link for LAN interconnection.

HIPERACCESS systems are means by which mainly residential customers and small- to medium-sized enterprises can gain access to broadband telecommunications and data communications services, delivered to their premises by radio. These systems are intended to be able to compete with and complement other broadband wired access systems including x (= generic) digital subscriber line (xDSL) and cable modems. In this context, broadband implies data rates of 2 Mb/s and above, and 25 Mb/s is the competitive benchmark for HIPERACCESS systems.

Thus, HIPERACCESS is a radio access system that may be deployed to connect subscriber terminals to a service provider node of a broadband core network, where the subscriber terminals are located in, and physically fixed to, customers' premises.

TABLE 8.7. Range of ETSI-BRAN Standards for Broadband Wireless Access

BRAN System	Frequency Band	Configuration	Application	Data Rate	Range
HIPERLAN 1	5 GHz (unlicensed)	MP-MP	Wireless LAN	20 Mb/s	50 m
HIPERLAN 2	5 GHz (unlicensed)	P-MP	ATM, IP, UMTS access	25 Mb/s	50 m
HIPERACCESS	>11 GHz (licensed)	P-MP	ATM, IP remote access	25 Mb/s	5 km
HIPERMAN	<11 GHz (licensed and unlicensed)	P-MP	IP local loop access	25 Mb/s	5 km
HIPERLINK	17 GHz (licensed)	P-P	LAN interconnect	155 Mb/s	150 m

General characteristics and design principles for a HIPERACCESS system include the following:

- Target users include residential households and/or typical small- to medium-sized enterprises (SME).
- The radio system is capable of providing users with a peak one-way (up-link or down-link) information rate of at least 25 Mb/s at the UNI, accessed by using appropriate standardized packet protocols such as ATM or IP.
- HIPERACCESS systems support both symmetric and asymmetric data flows, which may be duplex or simplex.
- The system uses multiple access methods in order to optimize the efficiency of spectrum utilization for bursty traffic.
- The radio system also efficiently supports legacy services, specifically POTS and ISDN (possibly over the native ATM or IP network service in a standardized way).
- HIPERACCESS systems can be deployed by public or private operators in a competitive regulatory environment in any of the frequency bands above 11 GHz for which they hold the license.
- HIPERACCESS systems will be capable of supporting the following user applications, though a specific implementation may support only a subset of these applications:
  - Internet and Intranet access
  - Home working
  - Video telephony and video conferencing
  - Real-time video and real-time audio
  - Computer gaming
  - Voice (using 64-kb/s and lower-bit-rate codecs)
  - Legacy voice-band modems (including fax)
  - Narrow-band ISDN

HIPERACCESS networks also incorporate strong security systems designed to prevent fraud and protect the privacy of the customers', service providers', and operators' traffic. Suitable network management functions for monitoring and maintaining system performance and QoS are implemented.

Similar to the IEEE 802.16 standard, the scope of the HIPERACCESS standard is also a specification for the radio interface. It primarily focuses on the physical (PHY) and data link control (DLC) layers which are core network and applications independent. In order to efficiently interface and interwork with different types of core networks, the specification also includes a convergence sublayer. The scope of the HIPERACCESS standard is illustrated in Figure 8.11.

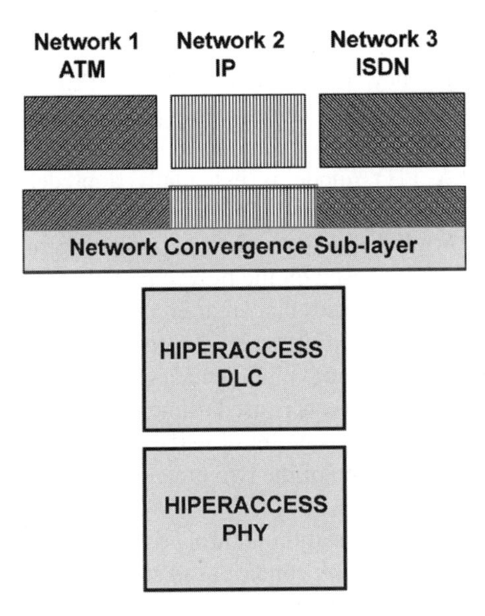

Figure 8.11. Scope of ETSI-BRAN HIPERACCESS radio specification.

8.6.3.2 HIPERACCESS Physical (PHY) Layer The architecture of the HIPERACCESS PHY layer is primarily driven by the following features:

- Single carrier (SC) transmission
- Support of multiple duplex methods
- Use of adaptive modulation and channel coding schemes

Duplexing Schemes In order to effect bi-directional, simultaneous transmission of information, HIPERACCESS system supports both FDD and TDD duplex operations. Additionally, it also supports as an option simplex or half-duplex using FDD (H-FDD) where simultaneous bi-directional operation is not required. In the case of FDD and H-FDD, the down-link and up-link channels are symmetric and have a bandwidth of 28 MHz each.

The H-FDD option leads to simpler and less costly subscriber terminals (referred to as access terminals or ATs in the ETSI-BRAN terminology). H-FDD is an optional feature as far as the ATs are concerned. The base station (referred as access point or AP in ETSI-BRAN specifications) is required to have the necessary capability to support H-FDD terminals.

In the case of TDD the same RF channel is time-shared between the down-link and up-link directions with no overlap between the time allocations between the transmissions in the opposite directions. For TDD operation, the access point (i.e., the base station) utilizes a frame-based transmission and allocates a portion of the frame for down-link transmission and the remainder for up-link transmission. The proportional allocation for the two directions can be varied on a frame-by-frame basis, thus providing different levels of asymmetry in bandwidth allocations in the two directions.

Modulation and Channel Coding Schemes (PHY Modes) In order to ensure maximum spectral efficiency and system range, HIPERACCESS specification adaptively sets the modulation and channel coding parameters (so-called PHY modes) according to the path loss or exposure to the amount of interference at individual subscriber terminals. A PHY mode is essentially a predefined combination of modulation and coding (FEC) scheme. Different PHY modes can be deployed in different parts of the down-link transmission frame, and different subscriber terminals can use different PHY modes based on the prevailing link conditions.

The range of modulation schemes that are used include QPSK, 16-QAM, and 64-QAM. Generally, the channel coding schemes consist of a Reed–Solomon (R–S) code that is concatenated with a convolution code (CC) in some of the PHY modes. The convolution code (when present) is used as the inner code, and the R–S code is used as the outer code. Typically, one set of mandatory and another set of optional PHY modes are available for each of the two directions of transmission. Each set in turn consists of three PHY modes.

The access point or the base station controls the assignment of a specific PHY mode. If the AP detects that the link conditions to some terminals have deteriorated, it will assign a more robust PHY mode to that link. When the link recovers from the interference problem, a less robust but more spectrum efficient PHY mode is assigned to the link. In some cases, only a limited number of the PHY modes may be deployed because of different, almost random interference behavior—especially under high levels of frequency reuse.

Conceptually, the functions provided by the transmitters and receivers implemented at the AP and the AT are similar. However, the equipment at the AP site is required to handle multiple RF channels and support multiple terminals and therefore requires an architecture that is different. From the PHY layer perspective, the basic functions performed at the transmitter include:

- Initial randomization of incoming data stream using a scrambler
- Channel coding (FEC) for error protection
- Preparation of the preamble bits (to precede a frame or TDMA region in down-link direction)
- Mapping of the data into symbols for modulation
- Pulse shaping
- Modulation and D/A conversion for radio transmission

At the receiver, the following functions are implemented to recover the user data from the analog signal received at the antenna:

- A/D conversion and burst demodulation to extract and identify the preamble
- Matched filtering operation to extract symbol values
- Equalization to enhance received signal quality
- Symbol de-mapping for translation to actual bit stream

- Channel decoding (FEC decoding) to assess data integrity
- De-scrambling to remove randomization performed at the transmitter before the data are submitted for continued processing in the user terminal.

8.6.3.3 HIPERACCESS Data Link Control (DLC) Layer HIPERACCESS DLC layer aims at providing a high level of spectral efficiency and multiplexing gain, at the same time ensuring required QoS over the links to satisfy the requirements of different traffic types. The DLC layer specification includes the medium access control (MAC) functions. The DLC layer for HIPERACCESS is connection-oriented, which means that MAC PDUs are received in the same order as they are sent and a connection is set up before the MAC PDUs are transmitted, thereby ensuring QoS. Key features of HIPERACCESS DLC layer include the following:

- It supports both FDD and TDD operation.
- It supports TDM and TDMA (optional) for down-link and TDMA for up-link.
- It operates in the connection-oriented mode (except for some up-link messages).
- It incorporates (AT-specific) radio resource control for load leveling, power leveling, and change of PHY modes.
- It incorporates (AT-specific) initializing control for initial access and release of terminals to/from the network as well as for reinitializing following link interruptions.
- It incorporates (connection-specific) DLC connection control for connection setup and release and connection aggregation.
- It incorporates ARQ protocol (using selective-repeat scheme) in the up-link channel to ensure QoS.
- It supports several request/grant schemes where requests are per connection basis or per connection aggregate basis and grants are per terminal basis.
- It supports privacy and security mechanisms for traffic privacy and fraud control.
- It uses fixed-size MAC PDUs to efficiently support ATM and IP.

Multiplexing and Frame Structures In the down-link direction, HIPERACCESS typically utilizes a time division multiplex (TDM) for transmission of data belonging to multiple ATs. As an option, a down-link frame may also carry time division multiple access (TDMA) signals. HIPERACCESS provides the capability to assign different PHY modes on a burst-by-burst (for TDM-based transmission) or time-slot-by-time-slot basis (for TDMA-based transmissions). Thus a down-link frame consists of a TDM region and (an optional) TDMA region. The TDM region contains of a number time segments, each of which is assigned its own PHY mode, and synchronization is provided once per frame for all ATs. In order to simplify the demodulation process, the PHY modes are assigned in a descending order of robustness so that an AT with worse link conditions (and assigned a more robust PHY mode) will end its reception process before an AT with better link conditions (and assigned a less robust but more spectrum-efficient PHY mode).

When a TDMA region is also present, an AT may receive information on the down-link in either the TDM region or the TDMA region as determined by the scheduler in the AP. The use of the TDMA option increases the overall channel utilization and reduces latencies. It is also necessary if the AP is required to serve half-duplex terminals. The down-link map broadcasts the allocations for both the TDM and TDMA regions of the down-link frame. The structure of the down-link frame is illustrated in Figure 8.12.

In the up-link direction, TDMA is used to support channel sharing between the subscriber terminals (access terminals) served by a base station (access point). The up-link bursts are scheduled by the AP after an AT is registered and makes a request for the channel. The scheduling data for individual ATs, consisting of time coordinates for start and end of transmissions, are organized in the up-link map, which is broadcasted at the beginning of the down-link frame as the broadcast control. Only registration messages and messages for bandwidth request are transmitted by an AT in an unsolicited contention mode.

8.6.3.4 HIPERACCESS Functional Elements (FEs) and Interfaces Figure 8.13 illustrates the major functional elements and interfaces associated with a HIPERACCESS system. As mentioned earlier, the specification essentially focuses on the radio interface (W1) and the other external interfaces consisting of W2 the service node interface (SNI) and W3 user–network interface (UNI) are typically selected from existing standard interfaces based on (a) specific implementations and (b) the services and applications they intend to support. The B3

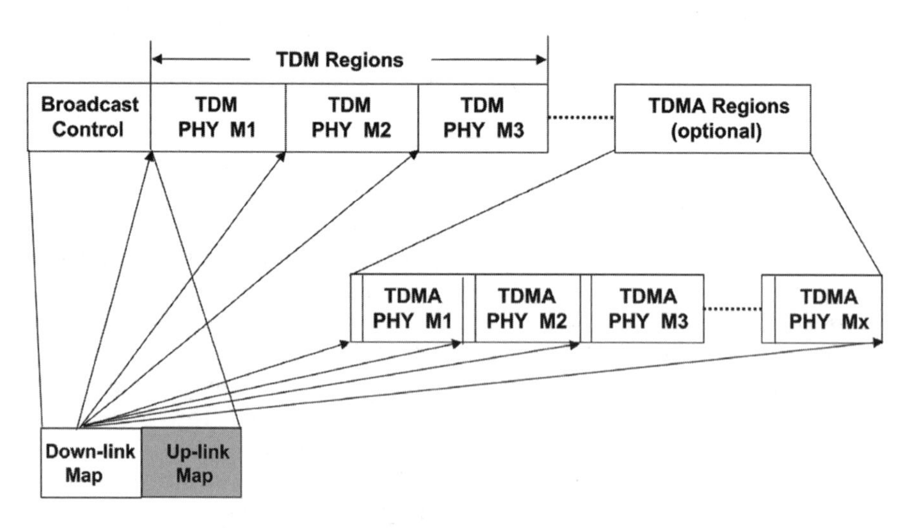

Figure 8.12. Down-link frame structure for ETSI-BRAN HIPERACCESS.

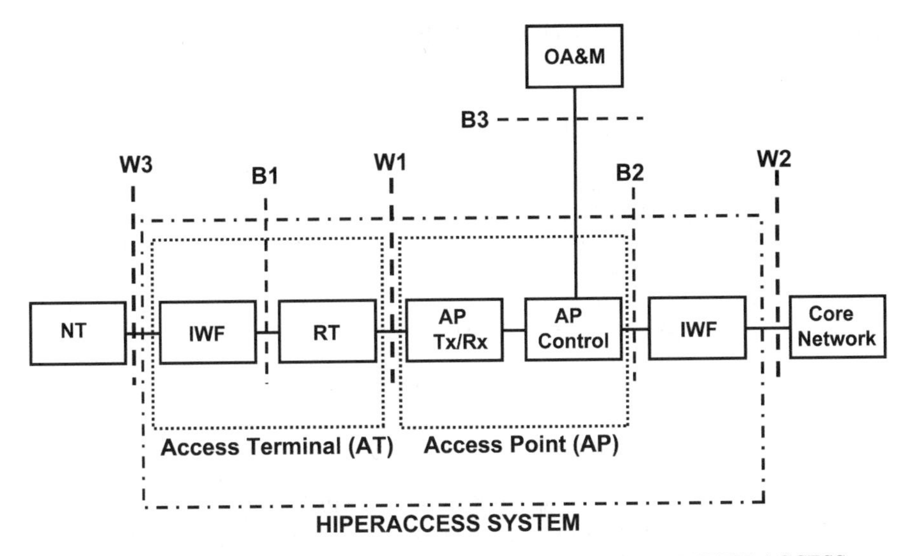

Figure 8.13. Functional elements and interfaces for ETSI-BRAN HIPERACCESS.

interface is the network management interface and may be based on appropriate TMN standards (e.g., ITU-T Recommendation M.3010). Interfaces B1 and B2 represent the interworking between the convergence layers defined in the specification and the external entities (network termination and the core network, respectively).

Some of the choices available for the UNI and SNI for supporting different services are summarized in Table 8.8.

8.7 BWA SYSTEMS BASED ON SATELLITE TECHNOLOGIES

8.7.1 Introduction

Satellite systems have been in operation for quite sometime for providing basic telephony services and for providing long-range regional and international routes. These systems generally use satellites located in the geostationary earth orbit (GEO) providing transponder capabilities. Use of on-board processing (OBP) and on-board switching (OBS) is generally not provided. Some satellite systems have also been developed to provide global mobility services to complement terrestrial cellular mobile networks. However, these systems have had little or very limited commercial success so far. Examples of such mobile satellite systems, sometimes referred to as global mobile personal communication using satellites (GMPCS), include Iridium, ICO, and Globalstar.

A number of satellite communication systems are currently being designed and developed to exploit the emerging market for high-quality multimedia information services on demand. These broadband satellite communication (SATCOM) systems

TABLE 8.8. UNI and SNI Options for HIPERACCESS Implementations

Service Type	User–Network Interface (W3) Options	Service Node Interface (W2) Options
PSTN	National standards for 64 kb/s, nx64, and T1/E1 connections	V5.1/V5.2 (ETS 300-347) VB5.1/VB5.2 (ETS 301-005)
N-ISDN	ISDN BRA (ITU-T Rec. I.430) ISDN PRA (ITU-T Rec. I.431)	V5.1/V5.2 (ETS 300-347) VB5.1/VB5.2 (ETS 301-005)
B-ISDN	B-ISDN UNI (ITU-T Rec. I.432)	VB5.1/VB5.2 (ETS 301-005)
ATM	ATM UNI (ATMF UNI 4.0)	B-ISDN NNI (ITU-T Rec. Q.2140/Q.2144)
	Ethernet CSMA/CD (ISO 8802.3) on 10 Base T or 100 Base T	ATM NNI (ATMF NNI spec.)
	DAVIC UNI (DAVIC 1.3)	FDDI (ISO 8802.5)
IP	Ethernet CSMA/CD (ISO 8802.3) on 10 Base T or 100 Base T	V5.1/V5.2 (ETS 300-347) VB5.1/VB5.2 (ETS 301-005)
	Serial Bus (IEEE 394 and USB)	ATM NNI (ATMF NNI spec.)
	ATM UNI (ATMF UNI 4.0)	B-ISDN NNI (ITU-T Rec.Q.2140/Q.2144)
	DAVIC UNI (DAVIC 3.1)	FDDI (ISO 8802.5)

will offer high-speed Internet access, interactive video distribution and multicasting services with a guaranteed quality of service (QoS) by service class. Some of the key features associated with satellite communication systems for broadband access include the following:

- Some of them intend to use multiple satellites (satellite constellations) in medium earth orbits (MEOs) or low earth orbits (LEOs) to counter the signal delay problems associated with geostationary earth orbit (GEO) systems.
- Frequencies used generally fall in the Ka band for communication with the satellite terminals and earth stations.
- Many of these systems deploy advanced capabilities such as on-board processing (OBP), on-board switching (OBS), and intersatellite links (ISLs).
- Gross capacity (bit rate) of these systems is in the Gb/s range over a very large coverage area.
- They support ISDN and IP/ATM-type networking capabilities.
- The target market for these systems is to provide high-bit-rate connectivity for small-to-medium enterprises (SMEs) and large Fortune 500 companies using small rooftop satellite terminals.

The critical advantage of these satellite-based broadband wireless access systems over their terrestrial counterparts is that satellite-based BWA systems can offer very high capacity over a wide coverage area that includes not only major population centers but also highly isolated locations like oil and gas operations and maritime routes. However, unlike terrestrial BWA systems where the initial capital outlay is moderate and can be increased incrementally as the customer base increases, satellite-based BWA systems require relatively very high initial and ongoing capital expenditure for launching, managing, and replacing (in the case of LEO and MEO systems) of satellites. Thus, in spite of their technical viability, their commercial viability is still an open issue and is responsible for frequent changes in business models, system design, and target dates for commercial operation.

8.7.2 Example Topology and Potential Applications for a Satellite-Based BWA System

Satellite-based BWA systems are being designed and developed using GEO, MEO, or LEO satellite systems. Typically, they deploy on-board processing (OBP) and on-board switching (OBS) and/or on-board routing (OBR). Use of intersatellite links (ISLs) in order to avoid multiple satellite hops (and associated increase in signal delays) is also utilized in some systems. The topology for a satellite-based BWA system illustrated in Figure 8.14 assumes a system using multiple LEO or MEO satellites with ISLs.

Coverage is the primary advantage of GEO satellites over LEO or MEO systems which makes GEO satellites well-suited for multicasting (of video programs). However, the higher orbit means that the round-trip delay is much larger and makes them less suitable for voice or interactive video applications. Higher altitudes also mean that the cost of launching individual GEO satellites is much higher than a LEO or MEO satellite. However, for a given coverage area, many more LEO satellites are required than GEO satellites, and therefore the total costs for the LEO constellation may be higher. Some of the parameters associated with GEO, MEO, and LEO systems are summarized in Table 8.9.

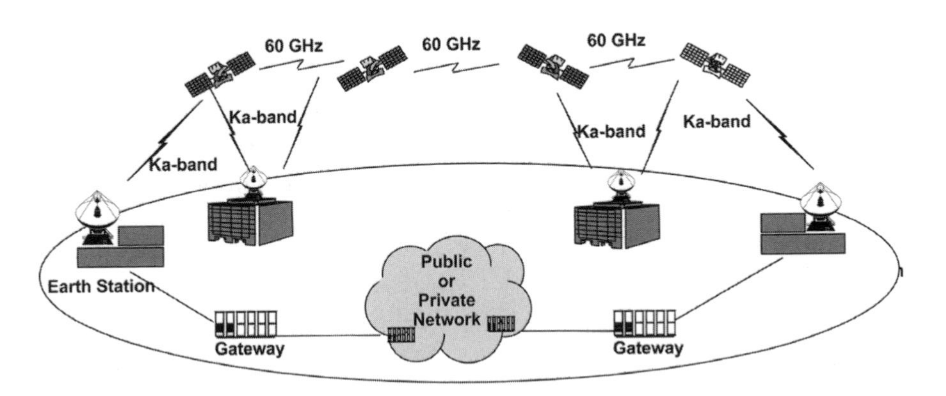

Figure 8.14. Example topology for a satellite-based BWA system.

TABLE 8.9. Characteristics of Different Satellite Configurations

Parameter	GEO	MEO	LEO
Altitude range	36,000 km	10,000–20,000 km	500–2000 km
Satellite visibility	24 h	2–4 h	10–20 min
Round-trip delay	500 ms	40–80 ms	5–10 ms
Satellite lifetime	20–30 y	10–15 y	4–8 y
Satellite constellation cost	Low	Medium	High

Though use of nongeostationary orbit satellite systems reduces the signal delay associated with GEO systems, it introduces some additional complexity in the design and operation of NGSO architectures. These include the following:

- The need for intersatellite handoff caused by relatively small footprint associated with NGSOs. As perceived by the satellite terminal, an NGSO satellite currently serving the terminal will move across the sky and disappear over the horizon, and the terminal will no more be within its footprint. In order to ensure continuity of service to the terminal, the traffic being carried by the satellite will need to be handed off in real time to another satellite that is visible to the terminal.
- The life span of NGSO satellites (specially LEO) is relatively short and will require periodic replacement which entails significant ongoing costs.
- NGSO systems are subject to shadowing effects associated with lower orbits (loss of signal due to shadow cast by buildings, etc.).

The range of applications that are expected to be supported by satellite BWA systems include the following:

- Global networking for enterprise networks which will seamlessly link a corporation's geographically distributed branch offices, subsidiaries, and remote workers with high-speed access to the corporation's Intranet.
- Broadband Internet access and bandwidth on demand (BoD) services provided directly to business and organizations wishing to bypass terrestrial facilities.
- Two-way multimedia connectivity for video on demand (VoD) and video multicasting (some systems may deploy terrestrial links for return path).
- Provide high-bandwidth, low-latency connectivity for government and nongovernment organizations (NGOs) around the globe using compact satellite terminals.
- Provide secure, reliable, and fast connections for maritime and aeronautical applications, as well as for oil and gas installations in highly isolated locations.
- Provide basic telephony services to underserved isolated locations.
- Support emerging applications such as telemedicine, interactive distance education, video conferencing, supply chain networking, and so on.

8.7.3 Examples of Emerging Satellite-Based BWA Systems

A number of satellite systems for broadband wireless access are emerging with commercial operation either already in progress or expected in the very near future. Some of these are briefly described here.

8.7.3.1 Teledesic The system was primarily funded and driven by a private consortium. The architecture and configuration of the Teledesic system has undergone a number of changes since its initial concept, which envisaged a constellation of 840 LEO satellites orbiting in 24 planes with intersatellite links. The last design proposal consisted of 30 MEO satellites with the initial service starting with 12 satellites, with the remaining 18 being added subsequently to provide global coverage. The target for commercial service availability was year 2005.

The Teledesic system was awarded a license for 1-GHz spectrum within the ITU assigned spectrum for nongeostationary satellite orbit (NGSO) service systems in the Ka band (28.6–29.1 GHz for up-link and 18.8–19.3 GHz for down-link). It was designed to provide intersatellite links operating in the 60-GHz band. A variety of user terminals were to accommodate *on-demand* single channel rates from 128 kb/s to 100 Mb/s on the up-link and up to 720 Mb/s in the down-link.

At the time of writing, activity on the future development of Teledesic system has been suspended, and its commercial availability in the near future is uncertain.

8.7.3.2 Skybridge This satellite BWA system is being developed by Alcatel Space. Its proposed configuration consists of a constellation of 80 LEO satellites (known as a Walker constellation) in 20 planes, each with four satellites in an orbital inclination of 54 degrees with respect to the equator. The architecture will deploy spot beams with frequency reuse to enhance system capacity.

The Skybridge system will operate in the Ku band (12.75–14.5 GHz for up-link and 10.7–12.75 GHz for down-link) in order to take advantage of available technology in the Ku band. The satellites in the Skybridge system only perform a relay function (bent-pipe relay architecture) and do not have the level of intelligence and switching capabilities found in some other systems. It will support interactive multimedia services at data rates of up to 100 Mb/s, with the majority of services being IP-based.

The original Skybridge service availability target was year 2001. However, it was announced in March 2001 that, in the interim, limited service will be provided using a GEO satellite, and currently the development of the 80 satellite system seems to be on hold.

8.7.3.3 Spaceway This is the satellite BWA system from Hughes Networks which is proposing to use constellation of 16 GEO and 20 MEO satellites. It will provide on-board processing (OBP) and on-board routing (OBR) capability and route traffic dynamically on the satellite, allowing full mash connectivity. This design feature is expected to yield lower latency and faster response time. The first

Spaceway satellite launch is planned for the year 2003, and service availability is expected in North America by 2004.

Spaceway will operate in the Ka band and will provide down-link speeds of up to 50 Mb/s and up-link speeds of up to 16 Mb/s with bandwidth on-demand operation. It is planning to serve large businesses, small office/home office (SOHO) users, and telecommuters for supporting a variety of broadband applications including desktop video conferencing, telemedicine, and interactive distance learning.

8.7.3.4 Astrolink

The satellite BWA system is being developed by Lockheed Martin and will use a constellation of nine GEO satellites operating in the Ka band (28.35–28.8 GHz and 29.25–30.0 GHz for up-link and 19.7–20.2 GHz for down-link). Astrolink is expected to start commercial service in the United States in 2003 and in the rest of the world in 2004.

The Astrolink satellites will have OBP and OBS capability for increased efficiency and flexibility, respectively. Data rates of 20 Mb/s on the up-link and 155 Mb/s on the down-link are expected. Its market focus is providing high-speed

TABLE 8.10. Main Features Associated with Some Selected Satellite BWA Systems

Parameter	Teledesic[a]	Skybridge	Spaceway	Astrolink
Main Backers	McCaw/Gates/ Saudis	Alcatel	Hughes Networks	Lockheed Martin
Number of satellites	30 MEO	80 LEO	16 GEO + 20 MEO	9 GEO
Frequency (UL), GHz	28.6–29.1	12.75–14.5	32.0–33.0	28.35–28.8 229.25–30.0
Frequency (DL), GHz	18.8–19.3	10.7–12.75	22.55–23.55	19.7–20.2
Service availability	???	2003?	2004	2003–2004
Access method	MF-TDMA ATDMA	CDMA TDMA FDMA	FDMA TDMA	FDMA TDMA
Network interfaces	IP/ATM/ISDN	IP/ATM	IP/ATM/ ISDN/FR	IP/ATM/ISDN
Gross capacity	10 Gb/s	20 Gb/s	4.4 Gb/s	6.5 Gb/s
Principal applications	Internet access, high-quality voice, high-speed data, video	High-speed Internet access, interactive multimedia	High-speed Internet access, bandwidth on-demand, multimedia services	High-speed multimedia

[a]If and when the system will be commercially available is unknown.

multimedia and Internet connectivity. Use of relatively small-diameter antennas (90-cm dishes) makes it suitable for large mobile platforms.

Main features of the above satellite-based BWA systems are summarized in Table 8.10.

Some of the other satellite-based BWA systems that are under development and proposed for commercial service include Cyberstar (Loran, USA), Euro Skyway (Alenia Spazio, Italy), GESN (TRW, USA), and iSky (privately owned, USA).

BIBLIOGRAPHY

Books

N. J. Boucher, *The Cellular Radio Handbook*, fourth edition, Wiley, New York (2001).

Martin P. Clark, *Wireless Access Networks: Fixed Wireless Access and WLL Networks—Design and Operation*, John Wiley & Sons, New York (2000).

A. Mehrotra, *Cellular Radio—Analog and Digital Systems*, Artech House, Boston (1994).

M. Mouly and Marie-B. Pautet, *The GSM System for Mobile Communications*, M. Mouly et Marie B. Pautet Palaiseau, France (1992).

R. Pandya, *Mobile and Personal Communication Systems and Services*, IEEE Press, Piscataway, NJ (2000).

P. Stavroulakis, *Wireless Local Loops: Theory and Applications*, John Wiley & Sons, New York (2001).

W. Webb, *Introduction to Wireless Local Loop*, second edition: *Broadband and Narrowband Systems*, Artech House, Boston (2000).

Documents from International Standards Development Organizations (ITU-R, ITU-T)

ITU Handbook on Land Mobile (including Wireless Access), Volume 1: *Fixed Wireless Access*, second edition (2000).

ITU-R Recommendation F.699: Considerations in the development of criteria for sharing between the terrestrial fixed service and other services.

ITU-R Recommendation F.1093: Effects of multi-path propagation on the design and operation of line-of-sight digital radio relay systems.

ITU-R Recommendation F.1096: Method of calculating line-of-sight interference into radio-relay systems to account for terrain scattering.

ITU-R Recommendation F.1399: Wireless access terminology.

ITU-R Recommendation F.1400: Performance and availability requirements and objectives for fixed wireless access to public switched telephone network.

ITU-R Recommendation F.1401: Frequency bands for fixed wireless access systems and the identification methodology.

Introduction to WLLs. By Raj Pandya
ISBN 0-471-45132-0

ITU-R Recommendation F.1402: Frequency sharing criteria between a land mobile wireless access and a fixed wireless access system using the same equipment type as the mobile wireless access system.

ITU-R Recommendation 1490: General requirements for fixed wireless access (FWA) applications.

ITU-R Recommendation F.1499: Radio transmission systems for fixed broadband wireless access (BWA) based on cable modem standard.

ITU-R Recommendation P.525: Calculation of free space attenuation.

ITU-R Recommendation P.530: Propagation data and prediction methods required for the design of terrestrial line-of-sight systems.

ITU-R Recommendation M.1457: Detailed specification of the radio interfaces of International Mobile Telecommunications 2000 (IMT-2000).

ITU-R Temporary Document 8A-9B/TEMP/87-E: Technical characteristics of broadband fixed wireless access (FWA) systems conveying Internet Protocol (IP) packets or asynchronous transfer mode (ATM) cells, September 19, 2002.

ITU-T Recommendation Q.543: Digital exchange performance design objectives, ITU-T Blue Book, Volume VI, Facile VI.5, Geneva (1994).

ITU-T Recommendation G.965: V-interface at the digital local exchange (LE)—V5.2 interface (based on 2048 kb/s) for the support of access network (AN).

Documents from Regional Standards Development Organizations (ETSI, TIA/EIA, IEEE)

ETSI TR 101 845 V1.1.1: Fixed radio systems; technical information on RF interfaces applied by fixed radio systems including fixed wireless access (FWA).

ETSI ETR 056: Digital enhanced cordless telecommunications (DECT); system description document.

ETSI ETS 300 175: Digital enhanced cordless telecommunications (DECT); common interface (CI); Part 1: Overview.

ETSI ETR 341: Digital enhanced cordless telecommunications (DECT); global system for mobile communications (GSM); DECT/GSM inter-working profile (IWP); profile overview.

ETSI ETR 139: Radio equipment and systems (RES); radio in the local loop (RLL).

ETS 300 347: V Interface at the digital local exchange (LE); V5.2 interface for the support of access network (AN).

ETSI EN 300 748 V1.1.2, Digital video broadcasting (DVB); multipoint video distribution systems (MVDS) at 10 GHz and above.

ETSI TR 101 173 V1.1.1: Broadband radio access networks (BRAN); inventory of broadband radio technologies and techniques.

ETSI TR 102 003 V1.1.1: Broadband radio access networks (BRAN); HIPERACCESS system overview.

ETSI TR 101 177 V1.1.1: Broadband radio access networks (BRAN); requirements and architectures for broadband radio access networks (HIPERACCESS).

ETSI TR 101 856 V1.1.1: Broadband radio access networks (BRAN); functional requirements for fixed wireless access systems below 11 GHz: HIPERMAN.

TIA/EIA IS-54B: Cellular system dual-mode mobile station–base station compatibility standard.

TIA/EIA IS-136.1: 800 MHz TDMA cellular-radio interface—mobile station–base station compatibility–digital control channel.

TIA/EIA IS-95A: Mobile station–base station compatibility standard for dual-mode wideband spread spectrum cellular system.

IEEE 802.16 2002: IEEE standard for local and metropolitan area networks—Part 16: Air interface for broadband wireless access systems.

IEEE P802.16.a/D3-2001: Draft amendment to IEEE standard for local and metropolitan area networks—Part 16: Air interface for fixed broadband wireless access systems—medium access control modifications and additional physical layers specifications for 2–11 GHz, March 2002.

IEEE 802.16.2 2001: IEEE recommended practice for local and metropolitan networks—coexistence of fixed broadband wireless access systems.

Journal and Conference Publications

A. Dennis, What next for WLL, *CDMA World*, December 1999.

G. Davies and S. Carter, et al., Key technological and policy options for the telecommunications sector in Central and East Europe and former Soviet Union, European Bank for Reconstruction and Development (EBRD) report, March 1995, London.

M. Rahnema, Overview of the GSM system and protocol architecture, *IEEE Communications Magazine*, December 1992, pp. 92–100.

R. Singh, Channel assignment schemes and traffic capacity of mobile radio systems, *IEEE Vehicular Technology Conference*, 1981, pp. 281–284.

B. Sarikaya, Packet mode in wireless networks: Overview of transition to third generation, *IEEE Communications Magazine*, September 2000, pp. 164–172.

A. Furuskar, S. Mazur, et al., EDGE: Enhanced data rates for GSM and TDMA/IS-136 evolution, *IEEE Personal Communications*, June 1999, pp. 56–66.

P. Bender, P. Black, et al., CDMA/HDR: A bandwidth-efficient high-speed wireless data service for nomadic users, *IEEE Communications Magazine*, July 2000, pp. 70–78.

E. Esteves, The high data rate evolution of the cdma2000 cellular system, *Multiaccess, Mobility and Teletraffic for Wireless Communications*, Vol. 5, Kluwer Academic Publishers, Hingham, MA, (2000), pp. 61–72.

A. Brazeau, Multi-access networks, *Alcatel Telecommunications Review*, 4th quarter, 2000.

A. Nordbotten, LMDS systems and their applications, *IEEE Communications Magazine*, pp. 150–154, June 2000.

P. Mahonert, T. Saarineri, and Z. Shelby, Wireless Internet over LMDS: Architecture and experimental implementation, *IEEE Communications Magazine*, May 2001, pp. 126–132.

M. Danesh, J.-C. Zuniga, and F. Concillo, Fixed low-frequency broadband wireless access radio systems, *IEEE Magazine*, September 2001, pp. 134–138.

J. Haine, HIPERACCESS: An access system for the information age, *Electronics and Communication Engineering Journal*, October 1998, pp. 229–235.

C. Eklund, R. B. Marks, and K. L. Stanwood, IEEE Standard 802.16: A technical overview of the wireless MAN^{TM} air interface for broadband wireless access, *IEEE Communications Magazine*, June 2002, pp. 98–107.

J. Forserotu and R. Prasad, A survey of future broadband multimedia satellite systems, issues and trends, *IEEE Communications Magazine*, June 2000, pp. 128–133.

J. Blineau, M. Castellanet, and D. Verhurst, Satellite contributions to the Internet, *Alcatel Telecommunications Review*, 4th quarter, 2001.

Selected World Wide Web Sources

Organization	Website
ITU—Radio Communications Sector (ITU-R)	www.itu.int/itu-r
ITU—Telecommunication Standardization Sector (ITU-T)	www.itu.int/itu-t
European Telecommunication Standards Institute (ETSI)	www.etsi.org
Institution of Electrical and Electronics Engineers (IEEE)	www.ieee.org
Telecommunications Industries Association (TIA)	www.tia.org
CDMA Development Group (CDG)	www.cdg.org
Qualcomm	www.qualcomm.com
Alcatel S.A.	www.alcatel.com
LG Electronics—S. Korea	www.lge.com
Lucent Technologies	www.lucent.com
NEC Corporation—Japan	www.nec.com
Nortel Networks	www.nortelnetworks.com
Siemens	www.siemens.com
Loyd's Satellite Constellations	www.ee.surrey.ac.uk/personal/L.wood/

GLOSSARY

3GPP	Third-Generation Partnership Project

A

AAA	Authentication, authorization, and accounting
AAL	ATM adaptation layer
ABR	Available bit rate
ABSBH	Average busy season busy hour
AC	Alerting channel; authentication center
ACH	Access response channel
A/D	Analog to digital (conversion)
ADPCM	Adaptive differential pulse code modulation
ADS	Asynchronous data service
ADSL	Asymmetric digital subscriber line
AGC	Automatic gain control
AGCH	Access grant channel
AK	Authentication key
AKA	Authentication and key agreement
AM	Amplitude modulation
AMPS	Advanced mobile phone system
AMR	Adaptive multirate (coding)
AN	Access network
ANSI	American National Standards Institute
ANSI-41	American National Standards Institute—41 (a.k.a. TIA IS-41)
AP	Access point
APAP	Any-point-to-any-point (communication system)
ARIB	Association of Radio Industries and Businesses (Japan)
ARQ	Automatic repeat request
AT	Access terminal
ATDMA	Asynchronous time division multiple access

Introduction to WLLs. By Raj Pandya
ISBN 0-471-45132-0

ATIS	Alliance for Telecommunication Industry Standards (USA)
ATMF	Asynchronous Transfer Mode Forum
ATPC	Automatic transmit power control

B

BB	Base band
BCC	Bearer channel control
BCCH	Broadcast control channel
BCH	Bose–Chaudhri–Hocquenghem (encoding)
BE	Best effort
BER	Bit error rate
BHCA	Busy hour call attempts
B-ISDN	Broadband integrated services digital network
BoD	Bandwidth on demand
BPSK	Binary phase shift keying
BRA	Basic rate access
BRAN	Broadband radio access network (ETSI group)
BS	Base station
BSC	Base station controller
BSD	Base station distributor
BSM	Base station manager
BSMAP	Base station mobile application part
BSS	Base station system
BTA	Basic trading area
BTS	Base transceiver station
BW	Bandwidth
BWA	Broadband wireless access

C

CA	Collision avoidance
CABSINET	Cellular access to broadband services and interactive television
CAC	Channel access control
CAGR	Cumulative annual growth rate
CAI	Common air interface
CAMEL	Customized application for mobile network enhanced logic
CAS	Channel associated signaling
CATT	China Academy of Telecommunications Technology
CATU	Central access and transcoding unit
CATV	Community antenna television
CBC	Cyber block chaining

CBR	Constant bit rate
CBS	Compact base station
CC	Convolution code
CCC	Cordless cluster controller
CCCH	Common control channel
CCFP	Common control fixed part
CCH	Control channel
CCS	Common channel signaling; hundred call seconds (36 CCS = 1 erlang)
CD	Collision detection
CDCS	Continuous dynamic channel selection
CDG	CDMA development group
CDMA	Code division multiple access
CDMA 1XEV	CDMA 1X evolution
CDMA 1XEV-DO	CDMA 1XEV—data only
CDMA 1XEV-DV	CDMA 1XEV—data voice
CELP	Code excited linear prediction (encoding)
CEPT	Conference of European Post and Telecommunications
C/I (ratio)	Carrier-to-interference (ratio)
CID	Connection identity
CLID	Calling line identification
CLNP	Connectionless Network Protocol
CLPC	Closed-loop power control
CM	Connection management; call management
CMIP	Common Management Information Protocol
CN	Core network
CO	CENTRAL Office
CPE	Customer premises equipment
CRC	Cyclic redundancy check
CS	Cell station
CS	Convergence sublayer
CSAP	Channel access control sublayer service access point
CSMA/CA	Carrier sense multiple access with collision avoidance
CSPDN	Circuit-switched public data network
CTA	Cordless terminal adaptor
CTIA	Cellular Telecommunication Industries Association
CTRU	Central transceiver unit
CWTA	China Wireless Telecommunications Association

D

DAMPS	Digital advanced mobile phone system
DAVIC	Digital audio–visual council
dB	Decibel

dBm	Decibel relative to 1 mW
DCS1800	Digital cellular system 1800 MHz
DCC	Digital control channel
DCCH	Dedicated control channel
DCS	Dynamic channel selection
DCS 1800	Digital communication system at 1800 MHz (GSM for 1800-MHz band)
DCW	Data code word
DDI	Direct dialing in
DECT	Digital enhanced cordless telecommunications
DES	Data encryption standard
DIU	DECT interface unit; data interface unit
DIUC	Downlink interval usage code
DLC	Data link control; digital line carrier
DOCSIS	Data over cable system interface specification
DPRS	DECT packet radio service
DQPSK	Differential encoded quadrature phase shift keying
DRC	Data request channel
DS	Direct sequence
DSI	Digital speech interpolator
DSL	Digital subscriber line
DSMA	Data sense multiple access
DSP	Digital signal processing
DS-SS	Direct sequence—spread spectrum
DSU	Data service unit
DTMF	Dual-tone multifrequency
DTx	Discontinuous transmission
DVB	Digital video broadcasting (standard)

E

E1	European primary rate channel (2.08 Mb/s)
E-BCCH	Extended broadcast channel
EBRD	European Bank for Reconstruction and Development
EDGE	Enhanced data rates for global evolution
EFR	Enhanced full rate (coder)
EIA	Electronic Industries Association
EIR	Equipment identity register
EO	End office
ERP	Effective radiated power
ES	Earth station
ESN	Electronic serial number
ETS	European Telecommunication Standard

ETSI	European Telecommunications Standards Institute
EVRC	Enhanced variable rate codec

F

FA	Frequency assignment
FAC	Factory approval code
FACCH	Fast associated control channel
F-BCCH	Fast broadcast channel
Fc	Carrier frequency
FCC	Federal Communications Commission
FCCH	Frequency correction channel
FDD	Frequency division duplex
FDMA	Frequency division multiple access
FE	Functional element
FEC	Forward error correction
FFSK	Fast frequency shift keying
FH	Frequency hopping
FM	Frequency modulation
FP	Fixed part
FPLMTS	Future Public Land Mobile Telecommunication Systems (now IMT-2000)
FR	Frame relay
FSK	Frequency shift keying
FWA	Fixed wireless access

G

GAP	Generic access profile
GFSK	Gaussian frequency shift keying
GEO	Geostationary earth orbit (satellites)
GERAN	GSM/EDGE radio access network
GGSN	Gateway GPRS support node
GHz	Gigahertz
GIP	GSM interface profile
GMPCS	Global mobile personal communication by satellites
GMSC	Gateway mobile switching center
GMSK	Gaussian minimum shift keying
GMSS	Global mobile satellite system
GOS	Grade of service
GPC	Grants per connection
GPRS	General packet radio service
GPS	Global positioning system

GPSS	Grants per subscriber station
GSA	Geographic service area
GSM	Global system for mobile communications
GSN	GPRS support node
GTP	GPRS Tunnel Protocol
GUI	Graphical user interface
GW	Gateway

H

HDFS	High-density applications in the fixed service
HDLC	High-level data link control
HDR	High data rate
HDSL	High-speed digital subscriber line
H-GMSC	Home–gateway MSC
HGSN	Home GPRS support node
HIPERACCESS	High-performance access (for broadband services)
HIPERLAN	High-performance radio LAN
HIPERMAN	High-performance metropolitan network
HLR	Home location register
HPA	High-power amplifier
HS	Handset
HSCSD	High-speed circuit-switched data

I

IC	Integrated circuits
IDT	International digital trunk
IDU	Indoor unit
IE	Information element
IEC	International Electro-technical Commission
IEEE	Institute of Electrical and Electronic Engineers
IETF	Internet Engineering Task Force
i/f	Interface
IF	Intermediate frequency
IFRB	International Frequency Registration Board
IMEI	International mobile equipment identity
IMSI	International mobile subscriber identity
IMT-2000	International Mobile Telecommunications—2000 (formerly FPLMTS)
IN	Intelligent network
IP	Internet protocol
IS	Interim standard (TIA/ANSI)

ISC	International switching center
ISDN	Integrated services digital network
ISL	Intersatellite link
ISM	Industrial, scientific, and medical (frequency band in the USA)
ISO	International Standards Organization
ISP	Internet service provider
ISUP	ISDN user part
ITS	Intelligent telephone socket
ITU	International Telecommunications Union
ITU-D	ITU—Telecommunications Development Sector
ITU-R	ITU—Radio Communication Sector
ITU-T	ITU—Telecommunications Standardization Sector
IU	Interface unit
IWF	Interworking function

L

L2TP	Layer 2 tunneling protocol
LAN	Local area network
LAPD	Link access protocol for D channel
LAPR	Link access protocol for radio (CT2)
LBR	Low bit rate
LD-CELP	Low delay—code excited linear prediction (encoding)
LE	Local exchange
LEC	Local exchange carrier (USA)
LEO	Low earth orbit (satellites)
LID	Link identity
LLC	Logical link control; link layer control
LMCS	Local multipoint communication system
LMDS	Local multipoint distribution system
LMT	Local maintenance terminal
LNA	Low noise amplifier
LOS	Line of sight
LR	Location register
LSR	Linear shift register

M

MAC	Media access control
MAN	Metropolitan area network
MAP	Mobile application protocol
MC	Multicarrier

MDS	Multipoint distribution system
MDU	Multiple dwelling unit
ME	Managed entity
MEO	Medium earth orbit (satellites)
MHz	Megahertz
MIN	Mobile identification number
MM	Mobility management
MMAP	Multimedia access profile
MMDS	Multichannel multipoint distribution system
MMS	Multimedia message service
MOU	Memorandum of understanding
MPEG	Moving Pictures Expert Group (ISO/IEC)
MPLS	Multi-protocol label switching
MP–MP	Multipoint-to-multipoint
MRTR	Mobile radio transmission and reception
MS	Mobile station
MSC	Mobile switching center
MSIN	Mobile station identification number
MSISDN	Mobile station ISDN number
MSN	Multiple subscriber numbering
MSRN	Mobile Station Routing Number
MSS	Mobile satellite service
MT	Mobile terminal
MTA	Major trading area
MTBF	Mean time between failures
MTP	Message transfer part (SS7)
MTTR	Mean time to repair
MTX	Mobile telephone exchange
MVDS	Multipoint video distribution system
MWA	Multimedia wireless access
MWS	Multi-wall set

N

N.A.	North America
NAP	Network access point
NAS	Network access server
NCC	Network control center
NE	Network element
NGO	Nongovernmental organization
NGSO	Nongeostationary satellite orbit
NID	Network identification number
NIU	Network interface unit
NLOS	Non-line of sight

NM	Network management
NMC	Network management center
NMS	Network management system
NMT	Nordic mobile telecommunication
NMU	Network management unit
NSS	Network switching system
NNI	Network-to-network interface; node-to-node interface
nrtPS	Non-real-time polling service
nrtVBR	Non-real-time VBR
NT	Network termination
NTT	Nippon Telephone and Telegraph

O

OA&M	Operations administration and maintenance
OBP	On-board processing
OBR	On-board routing
OBS	On-board switching
ODU	Outdoor unit
OFDM	Orthogonal frequency division multiplex
OMC	Operations and maintenance center
OOK	On–off keying
OQPSK	Offset quadrature phase shift keying
OSI	Open system interconnection

P

PAP	Public access profile
PBX	Private branch exchange
PCH	Paging channel
PCIA	Personal Communications Industry Association
PCF	Packet control function
PCM	Pulse code modulation
PCMCIA	Personal Computer Memory Card International Association
PCS	Personal communication service
PCS1900	Personal communication system 1900 MHz
PCM	Pulse code modulation
PDA	Personal digital assistant
PDC	Personal digital cellular (Japanese digital cellular standard)
PDSN	Packet data service node
PDU	Protocol data unit
PHL	Physical layer
PHS	Personal handy-phone service (Japan)

PHY	Physical (layer)
PIN	Personal identification number
PKM	Privacy key management
PLMN	Public land mobile network
PM	Phase modulation
PN (code)	Pseudorandom noise (code)
POS	Point of sale (terminals)
POI	Point of Interconnection
PP	Portable part
PPP	Point-to-point protocol
PS	PHS station
PS	Physical slots
PSK	Phase shift keying
PSPDN	Packet-switched public data network
PSTN	Public-switched telephone network
PTM	Point-to-multipoint (communication system)
PTP	Point-to-point (communication system)
PTT	Post, telephone, and telegraph

Q

QAM	Quadrature amplitude modulation
QCELP	Quadrature code excited linear predictive (coder)
QM	Quadrature modulation
QPSK	Quadrature phase shift keying
QoS	Quality of service

R

RA	Rate adaption
RADIUS	Remote authentication dial in user service
RAN	Radio access network
RAND	RANDom number
RCC	Radio common carrier; radio control channel
RCF	Radio control function
RCR	Research and Development Center for Radio Systems (Japan)
RDC	Remote distribution cable
RF	Radio frequency
RFP	Radio fixed part
RFTR	Radio-frequency transmission and reception
RLAN	Radio local area network
RLC	Radio link control; remote line concentrator
RLL	Radio in the local loop

RLP	Radio link protocol
RNC	Radio network controller
RP	Radio part; radio port
RPCU	Remote power and connection unit
RPE-LTP	Residual pulse excitation–long-term prediction (vocoder)
RR	Radio resource (management)
R–S	Reed–Solomon encoding
RSL	Received signal level
RSSI	Received signal strength indicator
RSU	Remote switching unit
RT	Radio termination
rtPS	Real-time polling service (IEEE 802.16)
rtVBR	Real-time VBR
RTS	Request to send
RTT	Radio transmission technology
RTU	Remote transceiver unit

S

SAAL	Signaling for AAL (ATM adaptation layer)
SAC	Service access code
SACHH	Slow associated control channel
SAN	Satellite access node
SAP	Service access point
SATCOM	Satellite communication
SBC	System broadcast channel
S-BCCH	SMS broadcast channel
SC	Service center; single carrier; spreading code
SCC	Satellite control center
SCCH	Synchronization channel
SCCP	Signaling connection control part
SCH	Shared channel
SCP	Service control point
SDCCH	Standalone dedicated control channel
SDF	Service data function
SDMA	Space division multiple access
SDO	Standards development organization
SDP	Service data point
SDU	Signaling data unit
SG	Study group (ITU)
SGSN	Serving GPRS support node
SID	System identification number
SIM	Subscriber identity module
SINR	Signal-to-interference/noise ratio

SIR	Signal-to-interference ratio
SME	Small-to-medium enterprise
SMS	Short message service
SN	Service node; subscriber number
SNI	Service node interface
SNDCP	Sub-network-dependent convergence protocol
SNMP	Simple Network Management Protocol
SNR	Signal-to-noise ratio
SOHO	Small office/home office
SP	Service provider
SPM	Subscriber pulse metering
SREJ	Selective REJect
SRT	Subscriber radio terminal
SS	Subscriber station
SS7	Signaling system 7
SSG/IMT	Special Study Group/IMT-2000 (ITU-T)
STM	Subscriber transceiver unit
STM	Synchronous transmission mode
STRU	Subscriber's transceiver unit
SU	Subscriber unit
SV	Space vehicle (satellite)

T

T1	North American primary rate channel (1.55 Mb/s)
TAC	Type approval code
TACS	Total access communication system
TCAP	Transaction capability application part
TCH	Traffic channel
TCP	Transmission control protocol
TDD	Time division duplex
TDM	Time division multiplex
TDMA	Time division multiple access
TEB	Truncated exponential back-off
TIA	Telecommunications Industry Association (USA)
TMN	Telecommunication management network
TMSI	Temporary mobile subscriber identity
TP	Transfer Protocol
TPC	Turbo product code
TRAU	Trans-coding and rate adaption unit
TSB	Telecommunications Standardization Bureau (ITU)
TTA	Telecommunications Technology Association (South Korea)

TTC	Telecommunications Technical Committee (Japan)
TUP	Telephone user part

U

UDL	Universal data link
UGS	Unsolicited grant service
UHF	Ultrahigh frequency
UIUC	Up-link interval usage code
UMTS	Universal mobile telecommunication system (ETSI)
UNDP	United Nations Development Programme
UNI	User–network interface
U-NII	Unlicensed—National Information Infrastructure (spectrum for)
UPCH	User packet channel
UTRA	UMTS terrestrial radio access
UWCC	Universal Wireless Communications Consortium

V

VAD	Voice activity detection
VBR	Variable bit rate
VC	Virtual circuit/channel
VCI	Virtual circuit identity
VDSL	Very high-speed digital subscriber line
VHF	Very high frequency
VLR	Visitor location register
VMSC	Visited mobile switching center
VoIP	Voice-over IP
VoD	Video on demand
VP	Virtual path
VPN	Virtual private networking
VSAT	Very small aperture terminal (Satellite)
VSELP	Vector sum excited linear predictive (vocoder)

W

WAP	Wireless Application Protocol
WARC	World Administrative Radio Council
WB-DS	Wideband—direct sequence (CDMA)
WLAN	Wireless local area network
WLL	Wireless local loop

WMAN	Wireless metropolitan area network
WPBX	Wireless PBX
WRC	World Radiocommunication Conference
WS	Wallset
WWW	World Wide Web

X

xDSL	(x = generic) DSL

INDEX

Introduction to WLLs. By Raj Pandya
ISBN 0-471-45132-0

ABOUT THE AUTHOR

Raj Pandya holds a Master degree in Radio Physics and Electronics from the University of Calcutta, India, a Master degree in Electrical Engineering from the University of Toronto, Toronto, Canada, and a Ph.D. in Electrical Engineering from Carleton University, Ottawa, Canada.

Dr. Pandya is currently a consultant on mobile communications, with a primary focus on Network Standards for IMT-2000; the international standard for the third-generation (3G) mobile communication system specified by the International Telecommunications Union (ITU).

Dr. Pandya has been active in the telecommunications standards activities of the ITU since 1980, where he has chaired working groups on International Numbering Plans, ISDN Traffic Performance, Network Capabilities to support UPT, Signaling and Switching requirements for current and future mobile systems, and Traffic Engineering and Performance for mobile networks. He also spent two years as Director of Teletraffic Research Center at the University of Adelaide, Australia.

Dr. Pandya has taught undergraduate and graduate courses on communications engineering in India and Canada. Dr. Pandya worked in the Systems Engineering Division of Bell-Northern Research and Nortel Networks for over 20 years and is the author of *Mobile and Personal Communication Systems and Services* (IEEE Press, 2000).